ISBN 978-0-484-61822-9
PIBN 10056120

REPORT ON THE MARINE ZOOLOGY OF OKHAMANDAL IN KATTIAWAR

Dwarka Headland and Bay, from the South.

[Photo by Vaughn Kelit Member, Baroda.

OTHER ZOOLOGISTS

PART II

LONDON
WILLIAMS AND NORGATE
14 HENRIETTA STREET, COVENT GARDEN, W.
1916

Dwarka Headland and Bay

REPORT

TO THE GOVERNMENT OF BARODA ON THE

MARINE ZOOLOGY OF
OKHAMANDAL IN KATTIAWAR

BY

JAMES HORNELL, F.L.S.

Government Marine Biologist, Madras, and formerly Marine Biologist to the Government of Ceylon

WITH SUPPLEMENTARY REPORTS ON SPECIAL GROUPS BY OTHER ZOOLOGISTS

PART II

LONDON

WILLIAMS AND NORGATE

14, HENRIETTA STREET, COVENT GARDEN, W.

1916

PRINTED IN GREAT BRITAIN BY
RICHARD CLAY AND SONS, LIMITED,
BRUNSWICK STREET, STAMFORD STREET, S.E.,
AND BUNGAY, SUFFOLK.

10975

CONTENTS OF PART II

To facilitate reference, the list of reports contained in Part I is appended:—

CONTENTS OF PART I

INTRODUCTION

WHEN sending Part I. of these reports to the press, I hoped to be able to include in the second part—that which is now sent forth—reports upon the whole of the remaining portion of the collection made during my short visit of inspection to Okhamandal in December 1905 and January 1906. Unfortunately the outbreak of the Great German War has made this impossible, particularly in respect of the two most important of the remaining reports—those on the *Polychaeta* and the *Tunicata*. Indeed, I am doubtful if the report upon the Polychaet collection will ever be available, as the latter was in the hands of Professor Ehlers of Göttingen at the time war broke out.

In these circumstances, I decided that as I had in hand a number of valuable papers, including two long and important reports on the sponges, together with others on the Hydroida, Medusae and Polyzoa, as well as a lengthy one on the Indian Conch which has considerable economic importance on the shores of the Gulf of Kutch, it would be better to publish these at once, and relegate any further reports that may come to hand at a future date either to a supplementary issue or contribute them to some other Indian Zoological publication.

The remarkable faunistic richness of the Okhamandal coast is demonstrated most emphatically by the wonderful variety of sponge, hydroid and polyzoon life catalogued in the valuable papers contributed by Professor A. Dendy, F.R.S., and Miss Laura R. Thornely. In this respect it must be remembered that the collections were made single-handed in the face of many difficulties and within the limited period of five weeks, during which my attention was in the main directed to work of another nature, faunistic collection being of subsidiary importance.

The Government of His Highness the Maharaja Gaikwar have : f ain most generously undertaken to defray the whole cost of production of the present volume, thereby making the zoological world in general, and that of India in particular, their debtors for this generous and liberal expenditure.

That the work done during my inspection of the sea bottom around Okhamandal has borne useful fruit is evidenced by (a) the establishment of a highly remunerative Placuna pearl-fishery on the north-east coast, (b) the development of the local chank-fishery, and (c) the appointment of fishery officers charged to organise and develop these and other marine industries. With regard to the first-named, it is to be noted that within a year from the inception of the enquiry the attention drawn in my first report to the Government of Baroda on the fishery potentialities of this outlying portion of His Highness's dominions, resulted in a commencement being made in the exploitation of the large beds of window-pane oysters (*Placuna placenta*) which were found during my visit of enquiry lying to the east of Beyt Island and Poshetra peninsula. The beds first exploited were largely great deposits of many generations of the shells of this species which had lived and died successively in shallow water in this locality. The pearls that had formed during the life of these shells were still *in situ* between the valves and could be obtained with ease by washing the shells and sieving the muddy sediment. The pearls thus found, although duller in lustre than those obtained from live shells, readily found a remunerative market, and large profits, I am pleased to learn, have accrued to the Baroda Government. The method of working these deposits, employed to date, has been to lease the right to collect the shells and extract the pearls for an annual rental.

The amounts so obtained are as follows for the years named, viz :—

Year.						Revenue. Rs.
1907–08	...	...	...	...	...	...	216
1908–09	...	...	...	...	...	...	325
1909–10	...	...	...	...	...	...	251
1910–11	...	...	...	...	...	...	355
1911–12	...	...	...	...	...	...	4,000
1912–13	...	...	...	...	...	...	12,100
1913–14	...	...	...	...	...	...	25,300
1914–15	...	...	...	...	...	..	16,001
1915–16	...	...	...	...	...	...	11,600
		Total	...	...	...	Rs. 70,148

From what I know of the local conditions, I am strongly of opinion that the beds of dead Placuna are of very great extent and that the portion exploited to the present is an insignificant proportion of the whole. What is wanted is to devise means for the efficient working of beds lying in deeper water, where the collectors at present find it impossible to work. Largely on this account, I understand that of late the lessees have turned their attention to the beds of living Placuna, from which they are getting a good outturn of pearls.

The Baroda Government are very much alive to the importance of developing and safeguarding this industry. To this end they have appointed a promising young man to the charge of the local pearl-fishing industry, and he, after receiving instruction in general methods at my hands, is now stationed in Okhamandal to carry on experimental pearl-oyster culture and general fishery investigation. He is now engaged in studying the life-history details of the window-pane oyster as exhibited locally, and has laid down considerable numbers on different bottoms in several selected localities to test comparatively the suitability of the different locations and the conditions which conduce to the rapid production of pearls.

A recent letter received from Mr. S. R. Gupte, the officer charged with this work, records the interesting fact that he has found vast rubbish heaps on the mainland opposite Beyt composed of waste from some ancient and extensive chank-bangle factories. The fragments are largely the discarded apical and oral pieces cut off and thrown away in the sawing of shells for bangle working ; they are similar in appearance and form to those figured in my " Indian Conch Report " on Pl. VI. (fig. 3). With them are associated many broken fragments of unfinished bangles—the whole being typical waste from a shell-bangle factory. That the industry was conducted on a very extensive scale is evidenced by the great abundance of these remains ; the Government lessee of the present-day chank-fishery is now exploiting the deposit, presumably for lime burning, selling the material at Surat at the rate of Rs. 13 to Rs. 14 per khandi of 20 Bombay maunds (560 lb.). Strangely enough, the industry of shell-bangle manufacture has now entirely died out in this locality, not a single shell bangle being manufactured at the present day in Kattiawar. Owing to the war having put an end to the importation of glass bangles from Austria, the demand for shell bangles has greatly increased in Bengal ; the result is a great appreciation in the value of conch shells, amounting in the case of several qualities to as much as 100 per cent. increase. This has reacted favourably upon the various conch or chank shell fisheries, and that of Kattiawar has benefited with the rest.

It is appropriate to note here that substantial progress has been made in the development of the local chank-fishery since I first drew attention to the importance of this subject in my original recommendations. In former years barely 3,000 shells were fished annually, and most of these were disposed of to pilgrims visiting Beyt. The greater attention now given to the industry through the efforts of the Director of Commerce and Industry has already had gratifying results ; from a catch of 3,000 shells the production has now risen to 16,000 per annum, with promise of very considerable further increase. Direct relations have been established with some of the most important of the Calcutta chank merchants, whereby enhanced prices have been obtained to the profit alike of the Baroda Government and the lessee.

An interesting minor shore industry not touched upon in my first report is that of cuttle-bone collection. During the south-west monsoon the " bones " of cuttlefish are cast ashore in great quantity on the Okhamandal coast, and as these are in considerable request in European markets at remunerative rates, the export is considerable. Mr. Gupte informs me that about 300 maunds are exported to Bombay annually, say 75 cwt. at 28 lb. per maund. Seeing that recent quotations which I have received from a leading London firm are £2 16s. per cwt., c.i.f. London, for selected pieces from 8 to 12 inches long, and £1 8s. per cwt. for mixed sizes, by advice and assistance in the disposal of this product, the Department of Commerce and Industry should be able to render it more profitable to the collectors, thereby encouraging them to greater and more sustained efforts, and to the organisation of the trade upon systematic lines in place of the present haphazard and intermittent method of collection.

With the enlightened encouragement of a sympathetic central Government anxious to develop the marine resources of the State, I am confident that the fisheries of Okhamandal have a bright future, though it is likely that progress will be slow, owing partly to difficulties in regard to transport which render access to good markets costly and slow, and partly to the conservatism and religious scruples of a coastal population hitherto with little interest in fishery matters.

In conclusion I am glad to avail myself of this opportunity to thank the various specialists who have been so kind as to furnish reports upon the collections, for giving so freely and ungrudgingly of their time and knowledge to make their contributions exhaustive and complete.

To the Government of His Highness the Maharaja Gaikwar I desire to express my great appreciation of the honour done me by entrusting the original enquiry and the general charge of the production of this publication to my hands. I trust that their recognition of the value of pure science in the economic development of the fishery resources of their State may find ample recompense within the near future, both in the betterment of the condition of certain sections of the coastal population and in the continued increase of revenue returns derived from the State monopolies constituted by the pearl and chank fisheries of Okhamandal.

JAMES HORNELL.

Government Fisheries Station, Tuticorin,
July 10*th*, 1916.

ERRATUM

NOTE UPON THE IDENTIFICATION OF THE EDIBLE OYSTERS OF OKHAMANDAL.

THE rock oysters which densely crowd the surfaces of the rocks about half tide level at Poshetra Point were wrongly identified in Part I, pp. 22 and 23. This species appears really to be the *Ostrea cucullata* of Born and of Lamarck.

Its main distinguishing characters are as follows : Outline roughly oval ; the left valve extensively attached, the cavity deep and cup-shaped, with a sacciform extension into the hollow beak region of the hinge, which is moderately elongated in freely-grown individuals ; the edges of this valve have a distinct tendency to grow upwards. Externally the left valve is folded into deep ridges passing radially outwards from the hinge and ending in a sharply dentate edge which interlocks closely with the edge of the upper or right valve. The latter is flattened and opercular in form. The muscle scars are normally purplish-black in tint, sometimes brown, rarely white. Very characteristic is a row of closely-set elongated denticulations seen a short distance inwards from the margin on the inner surface of the upper valve ; these fit into a corresponding series of furrows in the lower valve. Externally the shell is tinted an opaque pinkish-purple. Internally it is white, margined with purple or black. The size is generally small, seldom exceeding three inches in length.

The true identity of the fine edible oyster found on the muddy bottom of Aramra Creek and named tentatively *O. cucullata* in Part I, p. 22, is somewhat doubtful. It is certainly not *O. cucullata*, as that is the name of the small rock-oyster above described. In general form and in habit it closely approximates to the common estuarine oyster of South India, a species which appears to be identical with or extremely closely related to *Ostrea virginiana*, Gmelin, the most abundant oyster of the Atlantic Coast of North America. The Kattiawar oyster differs from both these forms by the possession of white muscle scars, which in both *O. virginiana* and the South Indian form are consistently purplish black. It is the same species as is found in local abundance in the muddy creeks of the Sind and Kutch coasts. Its shell sometimes attains a remarkable size ; I have one valve 14 inches long. The thickness is also characteristically great. It will probably be found to be *Ostrea gryphoides*, Schlotheim, or a variety thereof.

Very little is known at present concerning the number and relationship of the different Ostreids of Indian waters, and their nomenclature is in a state of considerable

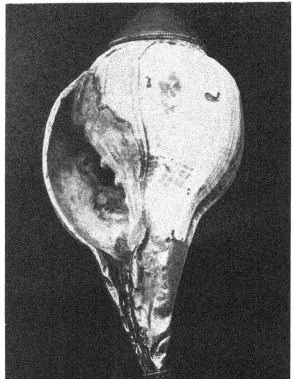

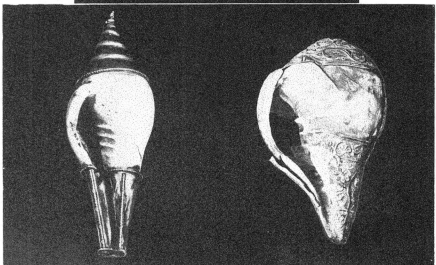

[Photographed by Vidulha Kala Mandir, Baroda.

Figs. 1, 2 and 3.—Sinistral chanks respectively in the Satya Bhamajı, Shank Narayan and Lakshmi Temples, Bet, Kathiawar.

THE INDIAN CONCH

(TURBINELLA PYRUM, *LINN*,)

AND ITS RELATION TO HINDU LIFE AND RELIGION

BY

.JAMES HORNELL, F.L.S.,

Marine Assistant, Department of Fisheries, Madras, and formerly Marine
Biologist to the Government of Ceylon .

[WITH SEVEN PLATES AND TWO TEXT-FIGURES]

INTRODUCTORY

To all Hindus the important position occupied by the Sacred Conch (*Turbinella pyrum*, Linn.), the *Sankha* of Sanscrit literature and the *Sanku* or *Chanku* of Tamil speech, in their religion is a commonplace of everyday knowledge ; few, however, are aware of the intimate relationship it bears to a hundred common incidents in the ordinary life of the people in many localities and among widely sundered races, tribes, and castes. In the following pages an attempt will be made to survey this little known by-way in the life and history of the Indian world, to show how superstition looks upon the Sankha as an amulet against the powers of evil, how this belief is among the oldest and most tenaciously held by Animist and Hindu, by Muhammadan and by Buddhist ; to indicate the way this belief has brought the shell into prominence in the Hindu religion, and to detail how it subserves as well, a hundred different uses in the daily lives of millions of Indians—how it is associated with infancy, marriage and death, and finally how its employment in the form of ornamental bangles was once the subject of an important industry in several widely separated parts of India, prominent being Kathiawar and Gujarat where to-day all memory of it has vanished.

In connection with this latter point I trust that the attention now drawn to it, may result in the revival in Kathiawar of an industry that has more good things than usual in its favour if we consider other aims than that of mere money-making ; no objects manufactured in India are more artistic and pleasing than the handsome milk-white bangles made in Dacca workshops for the ladies of Bengal.

In the course of collecting the materials for this essay, difficult problems have taken form. Some we can solve, but others remain obscure ; among the latter may be mentioned the unknown cause for the cessation and disappearance of the chank bangle industry in Kathiawar, Gujarat and the Deccan, and the question whether the use of chank bangles among a few sections of several castes in South India is in the nature of a survival of a once universal custom.

Another point to which I wish to draw attention is the bearing upon the antiquity of trade relations between India and the Persian Gulf in the recognition I made last year of several exhibits in the Louvre as consisting of objects carved from the shell of the Indian conch. True it is that they go back only some 500 years B.C. but it is something definitely accomplished when we can say with certainty that at this date we have positive evidence of the employment of the Indian chank, probably in religious service, in one of the great capitals of the Achæmenid dynasty of Persia. An examination of remains from Babylonian and Assyrian sites with a view to discover articles cut from this shell might well have profitable and important historical results.

I àm well aware of the many lacunæ in the story I offer, particularly in regard to ancient references to the Sankha preserved in the Sanscrit classics ; there may also be many valuable passages in ancient Gujarati works ; these were sealed books to me and I can but hope that others with the necessary qualifications may make good what is wanting in the present account.

[.

LIFE-HISTORY AND LOCAL RACES

THE Indian Conch, *Turbinella pyrum*, Linn., is a handsome gastropod mollusc found in Indian waters in large numbers in comparatively shallow water. Its geographical distribution is peculiar ; on the west coast of India large numbers are fished off the Kathiawar coast, but southward of this, we find no trace of the chank till we reach the southern coastline of Travancore where this shell again appears and forms the object of a small fishery. On the East Coast of India its distribution is more extensive, being found and fished everywhere from Cape Comorin to Madras City. The northern limit on this coast may be placed at the mouths of the Godaveri, where I have found a few shells, all marked by stunted growth—individuals living in an unfavourable environment. The northern shores of Ceylon, from Puttalam in the north-west to Trincomalee on the north-east, yield large numbers of this shell ; it is also to be found at the Andaman Islands.

The bottom most favoured of *Turbinella pyrum* is a sandy one containing a moderate proportion of mud ; this character of bottom is admirably suited to the luxuriant growth of tube-building polychæt worms which constitute the main food supply of the chank. These polychæts are of several genera, the most abundant being Terebellids. In some places in the Gulf of Mannar, square miles of sea-bottom are monopolised by these Terebellids and the chanks and echinoderms which prey upon them ; a veritable Tere-bella—Turbinella—Echinoderm formation. The edges of rocky reefs and the sandy patches interspersed among the rocks are other favourite haunts of chanks as the worms on which they prey are usually abundant there.

The shell of the adult chank is characteristically thick-walled and massive ; in the live condition the exterior is covered with a dense brown and velvety horny layer, the periostracum ; after death this dries, becomes brittle and eventually peels off, so that shells exposed for any length of time on the sea-shore or buried for any considerable period, become naked and reveal the characteristic snowy-white porcellaneous nature of the shell clearly. In fully grown shells the rows of chestnut brown spots which are normally very distinct on the outer surface in the immature, tend to disappear and often become entirely obliterated. Around the mouth, especially along the inner edge of the lip, a faint pink tint is often seen in fully mature shells, while occasionally the whole interior surface of the mouth may assume a brick-red colouration particularly in shells from certain localities. Characteristic of this genus are the three strong columellar plications or ridges upon the columella ; the trace of a fourth is also frequently present.

Local races.—So considerable are the divergences in external form and colour shown by shells of normal individuals of *Turbinella pyrum* that conchologists have made several pseudo-species out of the principal variations. Intimate acquaintance, extending over a period of upwards of ten years, with all the chank-producing localities on the coasts of India and Ceylon, together with the advantage of having scrutinised several hundred thousand shells, convince me that all the so-called species of Turbinella found in Indian seas are referable to a single species—*T. pyrum*—and constitute, at the most, varieties of the nature of emphasized local races.

Three types exist which may be conveniently described as the obtuse, the central and the elongate. All three forms have their own particular geographical distribution and characteristic physical and biological environment. Normally they respectively occupy well-defined localities. Intermediate forms linking the different main varieties frequently occur, especially where the habitats of two varieties march, forming a debatable land where it becomes difficult or even impossible to assign the shells there found to one or other variety, so evenly are the respective characteristics balanced. But the vast bulk of the shells fall readily into one or other of the three types.

(*a*). The central variety preponderates largely over the other two ; both numerically and industrially it is the most important and most valuable, being the form specially valued by the shell-bangle manufacturers of Bengal for the purposes of their trade. This type forms the bulk of the immense number fished off the Indian coast of the Gulf of Mannar and around the north of Ceylon, a number averaging some 1½ million annually. Its range on the Indian coast is straitly limited on the north, where Pamban Pass and the north coast of Rameswaram Island at the head of the Gulf of Mannar mark the boundary, thence southward it extends along the shores of the Ramnad and Tinnevelly Districts. North-westward of Pamban Pass this variety marches with that which I shall term *obtusa*. Off Ceylon this central form is restricted to the northern coasts from the Pearl Banks and Mannar round to Trincomalee on the north-east coast.

Figs. I and 2.—Immature chanks from Okhamandal; both
show persistence of the protoconch.

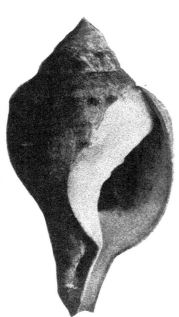

Fig. 3.—The Indian Chank, Turbinella pyrum, Linn., central
type of form.

Fig. 4.—Elongated variety from the
Andaman Islands. Adult and two
very young specimens. The latter
show the protoconch persisting.

[Figs. 1, 2, and 3 photographed by Vividha Kala Mandir, Baroda . Fig. 4 by Indian Museum, Calcutta.]

The central type has a heavy massive shell, with a well-proportioned and well-defined apical spiral. Apart from the chestnut coloured surface-markings which are extremely variable and tend to disappear with age, these shells are of an opalescent lily-white colour, porcellaneous in texture and slightly translucent when cut into sections. Unlike var. *obtusa* the colour of the body at the mouth is not deeply tinted ; usually it is of a very pale flesh tint or even quite white, whence comes the Tamil name *Vel-vayan* or " white-mouth " used to designate this variety, which again is split into several local races whereof the principal are known locally as (1) big-bellied chanks and (2) small-bellied chanks. The former come from the beds lying off the mouth of the Tambraparni River, the other chiefly from the northern beds off Tuticorin.

A second location of the central type is found on the coast of Kathiawar, separated from the Gulf of Mannar and Ceylon area by hundreds of miles of sea bottom from which all forms of *Turbinella pyrum* are absent.

(*b*). *Turbinella pyrum*, variety *obtusa*, is very distinctive in shape and sometimes in colouration. The apical spire is emphatically abbreviated, the whorls appearing as though telescoped while the colouring of the body at the mouth opening is frequently of a dark brick-red in shells from the Tanjore coast, most distinctive when compared with the delicate pallor of this region in shells of the central type.

The distribution of the *obtusa* variety includes the littoral waters of the eastern coast line of the Madras Presidency from Pamban Pass in Palk Bay to the delta of the Godaveri River in the north ; the chief fishing grounds are situated in Palk Bay and off the Tanjore coast, where this variety furnishes annually about 200,000 shells to the Calcutta market.

The shortening of the apical whorls often becomes extreme and so abbreviated as to be subtruncate and to approach in general form the aboral appearance of a typical *Conus*. This extreme form appears to arise when the environment is highly unfavourable, such as life in a muddy sea co-ordinated with scarcity of food ; it represents an extreme form of stunting, analogous to that seen on those pearl banks where dense over-population reduces the available food supply to a degree insufficient to meet the normal individual requirements.

(*c*). Variety *elongata* is a much elongated spindle-shaped form found in the Andaman Islands. As var. *obtusa* is the extreme in variation in the coniform direction so *elongata* approaches to the long-drawn-out spiral of *Fusus*. (Pl. II., fig. 4). Shells of this variety are very rare on the coast of the Indian mainland but. I cannot accept it as anything more than a very well-marked variety of *Turbinella pyrum*.

The anatomy of the body of *Turbinella pyrum* follows closely the lines of such a typical gastropod as *Buccinum* (Whelk), and it is unnecessary to go into any details here, save to mention that the sexes are separate, the proboscis much elongated, and the columellar muscles extremely powerful. It is well nigh an impossibility to extract the animal intact from its shell even after death ; the ordinary methods which are readily successful in the case of *Buccinum* are ineffective here, and the aid of a sledge hammer has to be evoked if we require to dissect the animal.

The ova are deposited in a many-chambered chitinous egg-capsule of striking and peculiar appearance. In general form it is an elongated loosely spiral annulated cylinder, divided transversely by partitions into a number of compartments ; it reminds one of a corrugated and loosely twisted ram's horn. When newly formed it is pale opaque yellow in colour ; with age it darkens and becomes covered with low growths of algæ. It stands upright on the sea-bottom, the lower and first formed end rooted in the sand by means of a broad flange-shaped basal disc. The lower end is narrow, the chambers there small ; these gradually increase in size, till at a point about one-third of the length from the base they attain a maximum size, which is maintained thence to the abruptly truncate summit.

Considerable complexity is introduced by the fact that the partitions between the chambers are double and separated by a slight space except along the hinder edge. The partition which forms the floor of each compartment has a crescentic slit parallel with the front edge, hence when the capsule sways to the current the partitions between the chambers gape slightly and so allow a circulation of sea-water within each compartment, thus providing the aeration needed by the larvæ. The transverse partitions are much thinner than the outer wall of the capsule.

In each chamber a considerable number of fertilised ova are deposited, embedded in a transparent colourless gelatinous nutritive matrix which entirely fills the chamber. In this nutritive jelly, the stronger of the embryos develop rapidly ; in each chamber an average of six reach the larval stage characterised by the possession of a larval external shell or protoconch ; those that do, grow rapidly and soon pass beyond the protoconch stage, and assume the semblance of adult form provided with a brown-flecked shell measuring eventually slightly over half an inch in length, inclusive of the two and a half whorls of the protoconch which persists both here and into adult life. By the time this size is attained all the nutritive contents of the incubatory chamber, including the remains of the weaker of the brethren, have been eaten up by the ravenous young who now find it necessary to leave home in search of food. Their first step is to eat through the partitions dividing the chambers, a proceeding that results in the bringing together of the whole surviving family numbering usually from 200 to 250 in all. The stronger next eat a way through the outer wall and the whole brood follow to scatter over the adjacent sea-bottom to lead independent lives.

The breeding season, when the capsules are fashioned and rooted in the sand, extends throughout January, February and the first half of March. Some divers assert that new capsules are also to be found in June and July, but I have had no opportunity to test this statement. The sexes being separate the divers have several stories to tell anent the breeding habits. Among others they assert that the females are the larger and are attended each by a number of smaller males, who assist in the making of the capsule—

a manifest impossibility. The post-larval rate of growth is quite unknown. On grounds where the food supply is abundant, it appears to be very rapid, for beds which are exhausted of adult shells measuring over 2¼ inches diameter in one year are found the succeeding year with plenty of shells, 3 inches and over in diameter.

The chank is an excellent instance of the acquisition by an animal of characters which appear for all practical purposes absolutely perfect to enable it to hold its own with ease in its struggle for existence. Against every one of its known enemies it has evolved suitable means of defence. The massive strength of the shell protects it from the attacks of all ordinary fishes ; the density and thickness of its periostracum give during youth and maturity adequate protection against the insidious attack of the boring sponge, *Cliona*, and its shell-burrowing congeners ; the strong capsule it constructs for its young gives them adequate protection till they reach a self-supporting stage endowed even at this early period with a fairly strong and resistant shell—one cannot crack it between one's finger and thumb. Its semi-burrowing habits give it great protection against those fishes which have the habit of snapping off the protruded feet of gastropods. Finally the pale yellowish-brown periostracum assimilates closely in colouring to the sand and should be a further protection against its discovery by foes ; to this form of protection I am, however, not inclined to assign great value, for chank divers can distinguish the presence of a chank even when half buried in the sand, and if they can, I feel assured that predatory fish are equally clever.

On rare occasions chank shells are found with the larger whorls crushed in ; this is believed to be the work of the great goggle-eyed Ray, *Rhinoptera adspersa*, which has the most powerful milling teeth of any Indian ray or shark.

As chanks grow old, their resisting powers diminish, the protecting periostracum receives damage and the burrowing sponge *Cliona* obtains a lodgment in the shell. Once there, it runs its branching tunnels everywhere in the substance of the shell, converting it into a honey-comb mass. I greatly doubt if this contributes except very occasionally to the death of the chank. This probably occurs usually from senile decay on beds that are not fished commercially. It is noteworthy that beds which have not been fished for some years, contain great numbers of Cliona-burrowed shells, whereas on beds fished regularly the proportion of " wormed " shells is so low as to be practically non-existent.

SOURCES OF INDUSTRIAL SUPPLY

EVERY year many lakhs of chank shells are required by the shell-bangle manufacturers of Dacca. To supply this demand hundreds of fishermen and divers, from Kathiawar in the north to Cape Comorin in the south, search the reefs and scour the sand-flats within the 10-fathom line during the fine weather season ; in some places they wade about in the shallows, in others they bring their catch to the surface in nets, or they may descend themselves to the bottom and hunt it by sight as they swim.

Six distinct chank fisheries are carried on at the present day in India seas ; ranked in their geographical order they are situated in the following localities :—

(a). Kathiawar ;

(b). Travancore ;

(c). The Gulf of Mannar (usually called the Tuticorin Fishery) ;

(d). Palk Strait (known generally as the Rameswaram and Ceylon Fisheries) ;

(e). The Coromandel Coast, from Point Calimere to Madras City.

Without exception the chank fishery in each of these localities is considered as a royal prerogative, the monopoly of Government. In practice this prerogative is variously exercised. In the Gulf of Mannar and on the Indian side of Palk Strait, the Madras Government work the fishery departmentally through an officer of the Fisheries Department styled the Superintendent of Pearl and Chank Fisheries. On the Coromandel coast the exclusive right to collect is farmed out to a renter for a term of years. The latter administration of the prerogative is also in force in Okhamandal (Kathiawar), where His Highness the Gaekwar of Baroda exercises sovereign rights in the local fishery. In Ceylon the renting system was in force till 1890, when it was abandoned in favour of an export duty, a method of securing Government revenue from this source which has continued ever since. In Travancore the dues of Government are collected in the same manner as now prevails in Ceylon.

The shells fished off the Kathiawar coast are of good quality, well esteemed in the Bengal trade where they are known as Sūrti shells—an echo of the day when Surat was the great emporium of the Kathiawar and Konkan coasts. To-day the shells are sent to Bombay, whence they are shipped to Calcutta. The quantity yielded is approximately 200 to 250 bags per annum.

Okhamandal, the north-western extremity of Kathiawar, which forms an outlying portion of the Gaekwar of Baroda's dominions, furnishes a considerable proportion of this export. The right to collect the shells is leased out at intervals for a term of years. Unlike other Indian chank fisheries the shells on this coast are all collected at spring tides when great areas of the littoral are uncovered at the time of low water. A certain proportion of the shells are sold to pilgrims who resort to the holy shrines at Bēt and Dwarka, the district of Okhamandal from its association with Krishna forming one of the holy lands of the Hindus, who delight to take home as a sacred souvenir one of the shells loved of this god. Full details of this fishery and of the enactments made to safeguard it, are to be found in Part I of the present Report.

Were diving for these shells to be introduced on the Kathiawar coast a great increase in the yield should be possible. As it is, the warm weather when alone diving can be systematically carried on is unfortunately that when the monsoon is most violent.

The chief centres of supply at the present day are the Gulf of Mannar and Palk Bay where physical and climatic conditions permit of diving being ultilised to an extent impossible in any of the other fisheries. Indeed chank-diving in these waters has been the regular calling of hundreds of the fishing population from time immemorial and references in Tamil classics make it clear that this fishery was prosecuted with vigour under Pandyan rule as long ago as the beginning of the Christian era ; in those days the headquarters of the industry was at Korkai, an important emporium with a population composed, we are told, in the main of traders, jewellers, pearl-fishers and chank divers. This city, the Kolkhoi of Greek geographers, was situated at the mouth of the Tambraparni which then entered the sea some 12 miles south of the present-day port of Tuticorin.

The relative importance of the various centres is shown clearly in the following table compiled from the Bengal customs returns for the eight years from 1905-1913.

TABLE SHOWING THE VALUE OF IMPORTS OF CHANK SHELLS INTO CALCUTTA FROM 1905 to 1913.

	1905-1906.	1906-1907.	1907-1908.	1908-1909.	1909-1910.	1910-1911.	1911-1912.	1912-1913.
	Rs.	Rs.	Rs.	Rs.	Rs.	Rs.	Rs.	Rs.
From Ceylon	144,772	189,280	86,515	181,223	166,060	87,716	133,495	127,940
From Madras—								
Chief Port	1,583	14,435	324	2,842	1,648	1,945	2,025	3,086
Other Ports	32,172	21,622	5,265	52,399	66,371	44,504	65,600	74,153
Travancore	114	—	592	—	500	—	96	—
Bombay*	?	?	?	?	?	?	?	?
Total ... Rs.	178,641	225,337	92,696	236,464	234,579	134,165	201,216	205,179

* Not ascertainable as chanks and cowries have been lumped together in the Customs returns of the Bombay exports.

a return, it does so with no niggard hand. The pearl fishery owes much to the chank-fishery; the latter is the school where boys and men of the fisher population pass their apprenticeship and qualify for the feverish days of a pearl fishery—a time when a gambler's lust for gain enters the diver's breast and makes him forget or ignore danger, and despise sleep and comfort in the race to gather the pearly treasures put at his disposal for an all-too-brief season, counted by days, and limited by the tempest.

III

THE ROLE PLAYED BY THE CHANK IN INDIAN RELIGION AND LIFE

(1). LEGENDARY AND HISTORICAL.

WHEN and how the cult of the chank as a religious symbol originated in India are questions going back so far beyond any traditions now existing that the utmost difficulty confronts us when we seek to find their solution. One main fact alone seems certain and that is the non-Aryan origin of this symbol. The Aryan-speaking hordes which descended upon the Punjab through the N.W. passes perhaps 2,000 years or more B.C., certainly did not bring the custom with them. They, the warrior ploughmen and herdsmen of the plains of Eastern Europe and Western Asia, had never seen the sea; they knew not as yet the deep sonorous boom of the snow-white chank—a note on a curved cattle-horn was with them the signal between scattered bands, while their hymns tell us that in music they used the drum, the flute and the lute. Vishnu, the God whose emblems include the chank, is barely mentioned in the Rig-Veda and the few Vedic hymns to him were probably composed after long intercourse had been established with the Dravidians, the chief race whom the invaders found in possession of the new land. He is almost certainly one of the gods borrowed from the indigenous people as his complexion is characteristically represented as dark-hued whenever his image is shown in colour.

I am strongly of opinion that the key to the problem is to be sought in the custom prevalent among animists—the worshippers of evil spirits—of employing noise to scare the demons they fear. At the present day Bengali Hindus make a practice of sounding the conch which each household keeps for religious rites whenever an eclipse or an earthquake occurs. Clearly this is a survival of the use of loud blasts on a shell to scare away evil spirits—the demon intent on devouring the moon or the sun, or shaking in fury the foundations of the world. This custom at once conferred religious significance upon the shell whose noise frightens the evil spirit and restores peace ; thenceforward the shell would be honoured and held sacred. Probably it was associated with one particular god, the prototype of Vishnu, and, with his adoption into the Aryan Pantheon, his emblem and weapon against the powers of evil would accompany him, and become endowed with still greater and deeper religious significance.

When the hungry swarms of Aryan tribesmen descended upon north-west India, the whole land, with the exception of the north-east corner, was occupied by a long-settled aboriginal population, split into many states and tribes vastly differing in civilisation. Many tribes, particularly those living in the mountains and dense forests and less accessible districts, were in the lowest possible stage, naked savages living on fruits and small game and maintaining a precarious defence against wild beasts by means of rude stone weapons and cudgels. In the south, especially in the maritime districts, a high civilisation developed at a comparatively early date and when the Aryan invaders were fighting their way into the Punjab against wild and semi-savage tribes, in appearance and customs probably much like the Santals 50 years ago, the men of the south were then or shortly later engaged in commercial relations with Babylon and the coastal districts of the Persian Gulf and Red Sea; partly through the stimulation received from this intercourse with these seats of ancient civilisation and partly from indigenous effort these southern Dravidians were evolving a language unsurpassed for its richness and flexibility and its power to express with perfect felicity the highest flights of imagination which poets and philosophers can reach, together with a material civilisation of no mean order. It is to these coastal Dravidians settled in the prosperous sea-ports situated on the western shore of the Gulf of Mannar or to men of the same race living on the Kathiawar coast that the first use of the chank must be traced. Both localities were the seats of pearl fisheries and the centres whence much oversea traffic flowed coastwise to Persia, Egypt and the adjacent Semitic lands. The chank and the pearl-oyster are usually associated in Indian waters, the chank on the sandy patches interspersed with the rocky patches which form the habitat of the pearl-oyster; pearl fishers often bring chanks ashore and thus the beauty of their snowy white porcelain-like massive shells would early become familiar to the merchants gathered from many lands to purchase pearls.

The word itself, in its chief forms of *chanku* or *sanku* in Tamil and *sankha* in Sanscrit, appears to be of Aryan origin; the Indian names are all obviously variants of one root and this is identifiable with the Greek *kongche* (κόγχη) and the Latin *concha*, both meaning a shell. It is probable that when the Aryans swarmed into India they applied their generic name for shells to the great white conch so conspicuous an object in the hands of their enemies, and so it is that, just as Christians in calling their scriptures by the Greek word for book (Biblos) imply thereby that it is " The Book " pre-eminent and supreme, so the Aryans appear to have similarly lauded the chank as " The Shell " and considered no qualifying term necessary. To them it was the shell of shells—the one shell above all others worthy of honour and even of worship. In the oldest Tamil literature the word is found in its present-day form but almost as often the shell is termed *vălai*; the latter is probably the original Tamil or Dravidian name, a term which has now come to be displaced by one derived from the Sanscrit *sankha*.

We may infer that when these ancient poems were written the Brahmans had already acquired great influence and were engaged in forcing Aryan forms and Sanscrit terms into the language of the Tamil country by virtue of their religious and literary ascendancy. As the religious leaders of the country, Sanscrit terms in constant use in their prayers and services would soon become dominant and be absorbed into the language of the country, in the way that Norman culinary terms readily found acceptance in the English language because of the superiority of the Normans to the Saxons in this particular art. Thus it has happened that the Sanscrit term for *Turbinella pyrum* gradually but surely ousted the indigenous Tamil term. Finally it is worthy of note that the English word " conch " is again another variant of the words chanku and sankha.

The earliest notices of the use of the chank are entirely of a secular nature, and occur in the two great Indian epics, the Ramayana and the Mahabharata. In these we get frequent reference to the employment of the chank as a martial trumpet by the great warriors whose more or less mythical exploits are recounted. Particularly is this the case in the Mahabharata, where in the Bhagavat-Gita we find the heroes endeavouring to terrify their foes with loud blasts on their battle-conchs, even as their forefathers scared away the evil-working spirits of the village. Each hero has his famous conch distinguished by some high-sounding name, just as the famous swords of European legendary heroes were frequently given names that have become immortal in song and story. The beautiful Excalibur wielded by Arthur in many glorious fights, Charlemagne's famous Joyeuse, and the magic Tyrfing so oft the theme of Viking sagas, have their parallels in the names of the conchs of the Mahabharata heroes.

When the opposing hosts of Kauravas and Pandavas confronted each other on the field of Kurukshetra, we read in the Bhagavat-Gita (verses 11 to 19) how the prelude to battle was the deafening clamour sounded by the leaders on their great conchs.

" The Ancient of the Kurus, the Grandsire (Bhisma), the glorious, sounded on high his conch, ' The Lion's Roar.'

" Then conchs and kettledrums, tabors and drums and cowhorns suddenly blared forth with tumultuous clamour.

" Stationed in their great war-chariot yoked to white horses, Mādhava (Krishna) and the son of Pāndu (Arjuna) blew their divine conchs.

" Panchajanya was blown by Hrishikisha (Krishna) and Devadatta by Dhananjaya (Arjuna). Vrikodara (Bhima) of terrible deeds blew his mighty conch, Paundra.

" The king Yudhishthira, the son of Kunti, blew Anantavijaya ; Nakula and Saha-deva blew their conchs Sughosha and Manipushpaka.

" And Kāshya of the great bow, and Shikhandi, the mighty carwarrior, Dristadyumna and Virata and Sātykai, the unconquered.

" Drupada and the Draupadeyas, O Lord of Earth, and Saubhadra, the mighty-armed, on all sides their several conchs blew.

" That tumultuous uproar rent the hearts of the sons of Dhritarashtra, filling the earth and sky with sound."

Here the names of the conchs possessed by all the five Pandava brothers are given,

—-Paundra, Devadatta, Anantavijaya, Sughosha and Manipushpaka. Paundra appears to have been named from Pundra, a demon killed by Bhima ; the others signify respectively God-given, Eternal Victory, Sweet Voice or Honey-tone, and Jewel-blossom.· Here too we hear for the first time the name Panchajanya given to the conch of Krishna, King of the Yādavas, who had espoused thé Pandavan cause. Around this famous shell many legends have gathered and now we see it held on high in most figures of Vishnu, who is considered by Hindus to have been re-incarnated in Krishna, the wise and good king of the Yādavas. According to one legend Panchajanya was originally the shell home of a terrible marine demon, Panchajana, so named as he was a foe to the five kinds of beings (*jana*), to wit, gods, men, gandharvas, serpents and ghosts or non-incorporated spirits. Panchajana lived on the sea bottom and at last filled the measure of his misdeeds by seizing the son of Sandipani, who had taught Krishna the use of arms. The God, fearless of consequences, rushed to the help of the child, assuming the form of a fish ; after a terrible struggle he vanquished the demon and brought away his shell as a trophy, since accounted one of the emblems of Vishnu and Krishna.

Tod, the author of the famous *Annals of Rajasthan,* in his " Travels in Western India," published 1839, in describing his visit to Dwarka and its neighbourhood gives a variant of this story and as the passage is most interesting no apology is needed for its reproduction in full. Under date January 1, 1823, he writes, " Crossed over to the Pirates' isle, emphatically called Bate, or ' the island,' but in the classic traditions of the Hindu, Sankhodwara, or ' the door of the shell,' one of the most sacred spots of his faith. It was here that Crishna [1] or Kanya acted the part of the Pythian Apollo, and redeemed the sacred books, slaying his hydra foe, Takshac, who had purloined and concealed them in one of those gigantic shells whence the island has its name. The whole history of Kanya, or Crishna, who assumed the form of Vishnu, is allegorical, but neither devoid of interest nor incapable of solution. There is no part of their mythology more easy of illustration than this, which is allusive to the sectarian warfare carried on at this period between the new sect of Vishnuvites and the more ancient one of Budha. The races who supported the religion of Crishna are typified under his emblem, Garuda, or the eagle ; while their wily adversary, the Budhist, is figured by the Tacshac, Naga, or serpent, a denomination given to the races of northern origin, which at various periods overran India, and of which were Taksiles (the friend of Alexander, the site of whose capital is still preserved in the Memoirs of Baber) and the still more famed Tacshac Salivahan, the foe of Vikrama. In the legend of the Yadu (Yadava) prince, Crishna (himself a seceder from the faith of Budha Trivicrama to that of Vishnu, if not its founder), receiving the sacred volumes from his hydra foe at this

---

[1] Tod notes " Kanya, or Vishnu, resembles the sun-god of the Egyptians in name as well as symbols. Kan was one name of the sun in Egypt, and his eagle-head is a well-known type." With regard to the extract given in the text, it has to be remembered that Tod's mythological explanations are not always reliable.

remote point of Hinduism, as well as his first combat with him in the Jumna, we have but the continuance of the same sectarian warfare, in which Crishna was in this instance successful, driving them before him both in the north of India and here : thus, his title of *Rinchor* was given on his defeat by Jarasindha, the King of Magadha, of the heretical faith, and at length these religious and civil conflicts led to his death, and the dispersion of the Yadu race of which he was the chief support. These Yadus, I surmise to have been all originally Budhists, and of Indo-Getic origin, as their habits of polyandrism alone would almost demonstrate ; and when we find the best-informed of the Jains assuring us that Nemnat'h, the twenty-second Budha, was not only Yadu, but the near kinsman of Crishna, all doubt is at an end ; and I am strongly inclined to pronounce decidedly, what I have before only suggested, that the Yadus are the Yute, or ancient Getes of the Jaxartes, amongst whom, according to Professor Newmann from Chinese authorities, one of the Shamanean sages sprung, eight hundred years before Christ. Both Nemnat'h and Sham-nat'h have the same personal epithets, derived from their dark complexions, the first being familiarly called Arishta-Nemi, ' the black Nemi,' the other Sham and Crishna, both also meaning ' dark-coloured ' ; and when this is not only confirmed by tradition, but the shrine of Budha itself is yet preserved within that of Crishna at Dwarica we have no reason to question that his faith, prior to his own deification, was that of Budha."

Always is Krishna's chank represented as a sinistral abnormality, and legend pictures to the mind of the devout Hindu every shell of this rare form when alive as a marvellous production receiving the homage of thousands of chanks of ordinary form, which crowd around it on all sides. Another myth is related by Baldæus, the chaplain to the Dutch forces which wrested Ceylon from the Portuguese, to the effect that Garuda, the eagle, vehicle or attendant (almost certainly the hawk-headed deity of Assyria) of Vishnu flew in all haste to Brahma and brought to Krishna " the chianko or kink-horn twisted to the right." [1] Vishnu derives several of his alternative names from his chank symbol, as Chankapāni, the " chank-armed," and Chankamēnthi, the " chank-bearer."

Krishna, when represented as a herdsman under the form of Govinda or Gopala, usually bears a conch in his hand and possibly the origin of this may be sought in the use a herdsman may make of it to call together his scattered charge just as the shepherds of Corsica and Sardinia at the present day use a great Triton shell (*Tritonium noduliferum*) for a like purpose.

A curious and most significant fact is that the twenty-second Tirthankar of the Jains, Nemi or Nemnath, who, legend says, was the son of King Samudravijaya of the race

---

[1] With the contradiction which exists between East and West in so many matters, the abnormal twist in these shells is termed the left-handed or sinistral by Europeans, whereas Indians term it right-handed. They view it from the mouth end, we from the apex, and accordingly confusion is frequent in conversation on this subject with Indians.

of Harivansa and a cousin to Krishna, has a conch as his emblem and is represented in Jain statues as being of a black colour. The black image of Nemi in the Nemnath temple on Mount Girnar in Kathiawar is a well-known example. The dark hue under which Vishnu and Krishna are always represented by Hindus and the black colour of his cousin Nemi, the Jain Tirthankar, go far to show that these gods and teachers belonged not to the Aryan race but to nations of Dravidian origin in the forefront of the earliest indigenous civilization in North India or Hindustan. With them in particular, is the conch most definitely associated; there is strong presumption on this and other grounds already referred to, to believe that it was the Dravidians who first employed the chank as a battle-conch and that this custom was adopted by the Aryan invaders as blood connections began to be formed in increasing numbers with the Dravidian nobility of the land and when certain of the Dravidian gods were admitted to the Aryan pantheon. The Aryans would be particularly eager to acquire fine conchs both for use and ornament ; their deep-voiced boom would prove their utility as battle-trumpets to enspirit and to give signals, while their rare white beauty would appeal to the religious sense as making them fit vessels wherewith to offer libations to their gods. To an inland people the beautiful products of the sea assume a double value from their strangeness and rarity and mysterious origin. To-day the people of Tibet, cut off from all knowledge of the sea, esteem pearls and red coral, tortoise-shell and amber, among the greatest treasures within their knowledge. The wild Nagas of the Assam hills equally prized the snow-white chank shell itself till some 50 years ago, using it as part of their accepted currency at the rates enumerated on page 43. As the Aryan hosts advanced into India they must have captured numbers of battle conchs from time to time and there can be no doubt they early adopted them in place of their own less sonorous cow-horns. Indeed the boom of the conch has been the battle signal throughout the ages in India, and this custom has lasted almost to the present day. Ancient Tamil and Rajput poems descriptive of battles and raids continually refer to the clamour of the conchs blown as the opposing parties approached each other ; the etiquette of old Indian chivalry required a prelude of challenging conch-blowing before the serious fight was begun ; the long-drawn hollow sonorous note of the chank often greeted early British commanders as they led their forces to the assault; until the beginning of last century Marathi and Pindari chiefs called their followers together and heartened them for the fray by loud blasts on conch-horns. Even in very recent days the chank's voice has called our enemies to the attack, and this too by other foes than Hindus. The graphic pen of Percival Landon in his " Lhasa "—an account of the British Mission to Thibet in 1903-4, in describing a night cannonade of the British Commissioner's post outside of Gyantse by the Thibetans, paints a word picture worthy of quotation:—" As one peers out into the warm night a long monotone is faintly droned out from the darkness ahead. It is one of the huge conch shells in the jong and it may only mean a call to prayer—the ' hours ' of Lamaism are unending—but as the moaning note persists softly and steadily a vivid speck of flame stabs the darkness across the river. A second later the report of the gun accompanies a prolonged 'the-é-es' overhead."

From the earliest times the conch has also been used in India to call the people to their sacrifices and other religious rites, as an instrument of invocation to call the attention of the gods to the ceremonies to be performed, and as a means to keep evil spirits at a distance and to prevent their entry into houses and temples.

With this intimate association with the chief religious rites, the people gradually came to reverence the instrument itself, and to adore and invoke it (see p. 22 for details), as is also done with regard to many other instruments or articles of sacrifice in Hindu rites at the present day ; these latter do not, however, appeal in equal measure to the religious feelings of the multitude, for around them have not been woven the myths and legends pertaining to the chank.

In the ceremonies attending the coronation of great kings the chank naturally played an important part. At the time when the Mahabharata was put into its present form, this custom was fully developed to judge from the description of the coronation of King Yudhistira given in the Shantiparwa of that epic. To quote from an interesting summary by Rao Sahib P. B. Joshi[1] "Kings of different countries, learned Brahmans and sages were invited for the ceremony. A *Vedi* or sacrificial altar was prepared. There were brought *Samidhas* or pieces of sacred wood, five kinds of sacred leaves, waters of the holy rivers and of the four seas, seven kinds of holy earth, the sacred conch shell, a white umbrella, and white *Chamars*. The horses and elephants used in connection with the coronation ceremony were also white. Yudhistir was then made to sit on a throne of gold, and other members of the royal family sat on seats made of ivory, and close by sat the king's spiritual guide and other sages. The king was now requested to touch such auspicious articles as corn, white flowers, swastika, gold, silver and jewels. The king's ministers and other high functionaries were now brought before him by the royal priest and they all paid their homage to their sovereign. The sacred fire was then kindled, the king and queen sat in front of the holy fire on seats covered over with tiger's skin, and made offerings to the fire. Shri Krishna then got up, took in his hand the sacred conch-shell, which was filled with holy water, sprinkled the water over the heads of the king and queen and said : ' I sprinkle this holy water over you to indicate that from this day you have become the paramount sovereign of Bharat Varsha.' At this time *dhundubi* and other musical instruments were sounded, drums were beaten, the king's bards sang the praise of the king and wished him victory and long life."

In Bengal every marriage conducted according to Hindu ceremonial includes the placing of chank-bangles, lacquered red, upon the bride's wrists. An iron bangle placed on the left wrist is also essential to the ceremony. Elsewhere this particular marriage custom is scarcely ever practised, but sufficient instances are recorded among a few widely scattered castes and caste-sections of the existence of a similar marriage custom

[1] *The Times of India Illustrated Weekly*, 20th September, 1911.

elsewhere to tempt us to believe that at one time it was the general custom of all fully Hinduised castes throughout India. Finds of fragments of chank-bangles in places where the use of these ornaments is non-existent at the present day strengthen the theory. Legendary lore can also be quoted in support. For instance, among the Balijas of Telugu districts, who there constitute the chief trading caste, a legend is current (Thurston, I, p. 137) that " on one occasion Siva wanted his consort Parvati to appear before him in all her glory. But, when she stood before him, fully decorated, he laughed and said that she was not as charming as she might be. On this, she prayed that Siva would help her to become so. From his braid of hair Siva created a being who descended on the earth, bearing a number of bangles and turmeric paste, with which Parvati adorned herself. Siva, being greatly pleased with her appearance, told her to look at herself in a looking-glass. The being who brought the bangles is believed to have been the ancestor of the Gāzula Balijas."

The latter sub-division of the Balijas peddle glass bangles only at the present day, but it is reasonable to suppose that before the discovery of glass, their stock in trade consisted instead of chank bangles. It is indeed probable that the introduction of glass dealt a heavy blow to the employment of the chank shell in feminine adornment in certain districts, particularly, for instance, in those where, as in Vizagapatam, glass factories being established, glass bangles were put on sale at a fraction of the cost of the comparatively expensive chank ones, which require the expenditure of much time and labour to render them attractive.

Another legend, prevalent among the Sangukatti Idaiyans, the great pastoral or shepherd caste of Tamil India, narrates that when Krishna desired to marry Rukmani, her family insisted on marrying her to Sishupalen. When the wedding was about to take place, Krishna carried off Rukmani and placed a bangle made of chank-shell on her wrist (Thurston, II, p. 354). These particular Idaiyans belong to one of the sections of this caste which to-day require their married women to wear these bangles—now a very rare custom in South India.

Indian sources give the barest indications of the traffic in chank shells that must have been brisk for 3,000 years or more between the fisheries in the Gulf of Mannar and on the Kathiawar coast and the inland nations of the Deccan and Hindustan.

(2) PRESENT DAY USES.

(a) IN RELIGIOUS CEREMONIAL (INCLUDING MARRIAGE AND DEATH RITES), AND VULGAR SUPERSTITION.

We have already seen that the chank is one of the two most important symbols— the other being the wheel or chakram—associated by Hindus with Vishnu, the second person in the Brahmanic trinity or Trimurthi. The majority of the avatars or incarnations of Vishnu are also occasionally represented as holding a chank in the hand : Matsya, in the form of a fish, Kurma the tortoise, Varaha the boar, and Narsingha the man-lion, are avatars sometimes sculptured holding Vishnu's chank : more frequently Krishna is thus depicted. Narayana, the god dwelling in the sun, another form of Vishnu, is similarly represented in human form with a chank in one hand and a discus (chakram) in the other. On rare occasions Siva is also depicted as holding a chank in one hand. In all these instances the chank represented is of the sinistral or left-handed form, a rarity so choice and valuable as to be worthy to form an adornment of a god. No more fitting gift to a deity can be imagined ; as the symbol of the god who divides with Siva the worship of the Hindu world, as a production of nature so scarce as to appear once only in several millions of normally shaped shells and as an emblem of purity, could Hindu find more fitting offering at the shrine of his god ? Thus it is that the pious wealthy have from time to time dedicated these shells to favourite temples—particularly to those that are in high esteem at centres of Hindu pilgrimage.

Chanks are held in special veneration at those shrines where the cult of Krishna is the predominant feature of worship, and in those localities rendered sacred to Hindus because of their reputed association with this deity while sojourning on earth.

Chief among the former is the great religious centre of the Madhva sect at Udipi in South Canara. There around a temple dedicated to Krishna are disposed eight mutts, or religious colleges, whereof the priesthood is a peregrinating fraternity, who tour the country to confirm in their faith the scattered members of their community— a procedure which consists in branding with hot irons the deity's symbols of the chank and the wheel upon the arms of the faithful.

Usually the priests of seven mutts are on tour ; the head of one mutt takes charge in rotation of the Udipi temple and for a period of two years carries on the services and provides the funds necessary for the lavish charity dispensed daily by the temple.

The members of this sect have a great reputation for active piety and the priesthood of each mutt vie with those of the others in collecting precious vessels for use in the service of the deity. The sinistral chank is the principal of these, and so it comes about that in the eight mutts at Udipi is gathered an unique collection of these rarities. During a recent visit to this place the Swamiyar or high-priest was good enough to show me the four which belong to his mutt—the Puttigé mutt. Of these, two were handsomely mounted in richly chased gold, the others in silver. All were rather small in size and one appeared to be of the sub-fossil description frequently found buried in muddy deposits in shallow lagoons in the north of Ceylon. A census of the sinistral shells possessed by the whole eight mutts was not possible, but from the information received from the high priest there cannot be less than thirty of these shells in the town, allowing the low average of three to each mutt and six for those possessed by wealthy Hindus who are not of the priesthood.

After Udipi, probably the largest number of sinistral chanks are to be found in the

B 2

temples of the sacred land of Kathiawar, rich in its association with the later life of Krishna ; while in Bet in 1906, I found richly ornamented sinistral chanks in the Shank Narayan, Lakhsmi and Satya Bhamaji temples. (Plate I.) That in the last named is a particularly large and handsome shell, probably the finest sinistral chank in existence and consequently an almost priceless treasure. The shell possessed by the Shank Narayan temple is a small elongated specimen offered at the shrine some twenty years ago by a Bathia from Zanzibar ; that of the Laksmi temple is a short broad one of small size with handsome arabesque ornamentation on the mounting—it has been in the possession of the temple since Samvat 1890 (A.D. 1835). At Benares, temple treasures include similar examples, while in the south of India, where opportunities to obtain these shells are great, a considerable number of the temples of that devout land possess one or more. Those at Rameswaram, Trichendur and Madura may be instanced.

The temple at Rameswaram, one of the four most holy places of pilgrimage among Hindus, possesses four shells ; of these three are very ancient and date from the days when the munificence of the Madura and Ramnad rulers endowed this temple with a share of the produce of the local pearl and chank fisheries. The shells in the Trichendur Temple, a towering pile overlooking a sea where chank fishing still flourishes, are derived likewise from a privilege share possessed formerly by the temple in this fishery.

It is noteworthy that although Rama, an incarnation of Vishnu, is especially associated with the Rameswaram temple, the actual deity represented in the inmost shrine is the phallic lingam of Siva. On important religious festivals water is poured into the sinistral chanks, and after Vedic chants have been sung and prayers offered, the water thus consecrated is poured over the lingam.

It is remarkable also and indicative that this custom has not originated with modern Hinduism, that sinistral chanks are objects of adoration among northern Buddhists. Sarat Chandra Das, the intrepid survey officer who spent some perilous years in Thibet, mentions (*Journey to Lhasa*, London, 1902) that in the Sakya Monastery lying to the south-west of Shigatze there is preserved in the temple a chank of this rare form. Its history is invested with more than ordinary interest, for the monastic records state that it was a present from Kublai Khan, the great Tartar conqueror of China and patron of the Polos, to Phagpa, a hierarch of Sakya whom Kublai made ruler of Thibet in the second half of the thirteenth century. Sarat Chandra Das mentions that this famous shell is blown by the lamas only when the request is accompanied by a present of seven ounces of silver, but to have it blown " is held to be an act of great merit."

In Thibet these left-handed chanks are called *Ya chvil dung-kar* and in Chinese *Yu hsuan pai-lei*. The people of both countries consider such shells as treasures of inestimable value. In 1867, one was known to be kept at Fuchu by the Ti-tuh (*Peking Gazette*, February 23rd, 1867) and one at Lhasa (*fide* an editorial note in Das, " *Journey to Lhasa*," above quoted).

At one time the value of these shells is said to have been assessed at their weight in gold and this statement is probably correct. To-day they are less valuable and small and imperfect ones occasionally change hands in the north of Ceylon at Rs. 60 to Rs. 90 each (say £4 to £6) ; such shells are usually sub-fossil ones found buried in deep mud. Large good-conditioned sinistral shells obtained alive, and therefore of good colour, still command high rates—anything from Rs. 400 to Rs. 5,000 or more—so infrequently are they found.

The method of estimating the approximate value of these latter shells is as follows— if the shell be of 100 rupees' weight or over (one standard coin weighs exactly one tola ; 2½ tolas = one ounce) the value is calculated at the rate of Rs. 50 per tola or rupee's weight ; thus a shell of 110 rupees' weight would be estimated accordingly to this method at Rs. 5,500. When the weight is between 30 and 100 tolas then the rate per tola may be taken at Rs. 30/- to Rs. 40/- according to size and quality ; a 50 tola weight shell would be worth from Rs. 1,500 to Rs. 2,000. If of 25 tolas' weight the rate may vary from Rs. 5/- to Rs. 20/- per tola.

Wealthy Hindu Rajahs and Zamindars also aspire to possess these sacrosanct shells, the orthodox in order that they may use them to perform *abeyshekam* [1] in their family shrine and others for the ostentatious pride of possession and because of the superstition generally current which accounts the ownership of a Valampuri chank as conferring prosperity upon the household owning it.

I know also a wealthy Muhammadan Marakkayar who has refused offers of several hundreds of rupees for one of these shells ; to him the shell is a mascot, the bringer of good luck.

The result of my inquiries indicates that there exist at present in India not less than 120 sinistral chank shells. I know the actual whereabouts of 60 and I am well satisfied that it is improbable that I have overestimated the remainder at an equal number.

The chank shell is so massive and strong as to be practically indestructible under ordinary circumstances. I have found chank artifacts in the ruins of Korkai, the Kolkhoi of the old Greek and Egyptian geographers, which though over 1,000 years old are indistinguishable except in colour from fragments cut in Dacca workshops to-day. Hence the sinistral shells now found in Hindu temples should represent the aggregate catch of this abnormality for several hundreds of years. If we place the period at the conservative limit of 300 years we shall find the percentage borne by sinistral to normal dextral shells an extremely small one. Elsewhere (Madras Fisheries Bulletin, No. 7) I have shown that the average catch of chank shells off the Indian and Ceylon coasts aggregates 2,500,000 per annum. There is reason to believe that a hundred years ago the average was considerably higher but some allowance must be made for interruptions in the fishery due to political unrest in the troublous times prior to the coming of the *Pax Britannica*. 2½ millions of shells are therefore probably a fairly accurate approxi-

---

[1] The worship of the gods with libations.

mation.   At this rate the produce of the past 300 years would amount to a total of 750,000,000 shells ; the 120 sinistral shells believed to exist now thus give the infinitesimal ratio of one shell to 6¼ millions of normal ones, and this proportion undoubtedly gives a substantially accurate idea of the great rarity of the abnormal form.

Among the ignorant who usually have never even seen a Valampuri chank, the belief is prevalent in Tamil South India that it blows of its own accord during the night ; even the Roman Catholic chank divers of Tuticorin entertain this quaint superstition, and say that the shell is specially clamorous on Tuesday and Friday nights ! A yogi when controlling or retaining his breath is also credited with hearing the sound of a Valampuri chank blowing within his abdomen though why the rumbling should be that of a Valampuri and not of an ordinary chank is hard to answer.

Worship of the chank as one of the three essential articles used in domestic worship among Brahmans should occupy an important part in the daily liturgy of this priestly caste and in the Brahma Karma, the work which sets forth in minute detail the order and phrasing of the sacred rites of the Brahmans, the prayer to the sacred chank may be translated as follows :—

Taking the chank in his hand the Brahman recites : " At the mouth of this shell is the God of the Moon, on its sides is Varuna, on its back Prajāpati, and on its apex the Ganges, the Sarasvati, and all the other sacred rivers of the three worlds in which they make ablutions according to the command of Vāsudeva.[1]   In this chank is the chief of the Brahmans (Brahmendra or Brahmanaspati).   This is why we must worship the sacred chank.   Glory to thee, sacred shell, blessed by all the gods, born in the sea, and formerly held by Vishnu in his hand.   We adore the sacred chank and meditate upon it.   May we be filled with joy !

" I offer (to the chank) everything needful for worship—perfumes, rice and flowers."

Here they make the sign of the chank, but Bourquin (" Annales du Musée Guimet," Vol. VII, p. 45), from whom I quote, says he was never able to discover the manner of making this sign and I have had a like ill-success.

While the above is a portion of the liturgy which the head of each Brahman family is theoretically bound to recite daily, under present-day conditions this is impossible, and in fact it is only on specially important puja or holy days that even a greatly curtailed version of this and the accompanying prayers is recited by the head of the family, and this, too, only in thoroughly orthodox families.   The curtailed prayer usually runs, " Oh, chank shell, thou wast produced in the sea and art held by Vishnu in his hand.   Thou art worshipped by all the gods.   Receive my homage."

In this connection an interesting chank legend centres round the temple tank in Tirukalikundram, a holy Saivite village in the Chingleput District, some 37 miles south-ward of Madras.   The village munsiff, Mr. T. A. Vedachala Gurukkal, to whom I am in-debted for the following particulars, states that once every twelve years a chank rises

[1] One of the names of Krishna.

to the surface of the large sacred bathing tank called Sanku Theertham. Three days before this marvellous occurrence, the water in the tank is greatly agitated, foaming and boiling to the accompaniment of considerable noise. On the appearance of the chank the foam subsides and the sound ceases. Then the chank comes to the side where it is picked up, purified and holy water sprinkled upon it. Thus consecrated it is carried in solemn procession through the village to the Vethagiriswarar temple, where it is placed in the treasury with the rest of the temple treasures.

Locally this chank is considered superior to all other chanks in the world and the origin of the periodic miracle is narrated as follows :—

When the sage Märkkandēya was visiting the sacred places with his disciple, the latter forgot to bring the Siva puja-box containing the chank required for the libation necessary in the ritual of worship. The sage took his bath in the tank and as puja time was fast approaching he prayed to be helped in this misfortune. As a result of his prayer and by the special grace of the deity, a chank appeared in the tank. Then he placed a lingam before him and, with the help of the chank miraculously provided, performed his worship in an acceptable and proper manner. He also prayed that a chank might appear each day he should worship at this tank. The Puranas say that this request was granted and has been since continued until the present day. In this connection my informant remarks that it is to be noted that twelve ordinary years constitute one divine day.

Thousands of pilgrims resort to the Tirukalikundram shrines each year and the marvel of this story is one of the great assets of the place just as is the periodic lique-faction of blood to the shrine of San Gennario in Naples.

In temple worship, the chank fulfils important service. The ordinary and sinistral forms are both employed whenever the temple possesses them. The former is used in the menial duty of summoning the god's attention, announcing the commencement of the principal rites as well as in calling the devout to worship ; such are among the general explanations given for its employment, but some ethnologists hold that the innate and primitive significance of the use of the blowing chank in temple worship is to scare away hostile and evil-working spirits. This is a reasonable belief as there is little or no doubt that the chank was used originally as a horn or trumpet by tribes holding animistic beliefs prior to the development of the Brahman religion which appears to have adopted the use of the chank in religious ceremonies together with many other rites from the devil-fearing tribes who gradually came into the fold of the new and higher religious belief.

In this connection should be mentioned the custom which prevails largely in Bengal of keeping blowing chanks in the houses of the better class people for use in family worship. Mr. A. R. Banerji, I.C.S., the late Dewan of Cochin State, informs me that it is a general custom in Bengal to turn out with these shells during eclipses and earthquakes and to keep up a continuous blowing till the eclipse or earthquake be over.

A rather striking effect is produced when the chank is used in temple ritual as a sort of rhythmical accompaniment when it plays the part of kannagōlu or tālavinyāsa.[1]

In Hindu temples the four daily services take place before dawn, at noon, at sunset, and at 9 or 10 P.M. At these times the Ochchans, the caste entrusted solely in the Tamil country with the duty of chank-blowing in temples, announce the commencement of each service and punctuate the various rites with the drone of their shell according to the customary ritual. Among the Uriyas the corresponding caste of temple servants is that of the Rāvulos, whose caste duty is to sound the chank during services in Saivite temples when the god is being taken in procession as also to prepare garlands of flowers for the adornment of the god. Like the Ochchans, they are not usually whole-time servants of the temple, but while the former earn money as musicians at weddings, performing upon a long silver trumpet, the Rāvulos make and sell garlands to the laity. The Ochchans never use the chank except in the temples, whereas the Rāvulos are employed to sound it at Brahmans' weddings. It is a rule among the latter that they must possess at least two blowing chanks, lest, losing one, the temple service should suffer in consequence.

In Bengal, the Ramavat sect of Vaishnavas pay particular attention to the call of the chank. By them all forms of worship except the unceasing repetition of the name Rama or Hari are deemed useless, but in every ākhāra or monastery of the sect an idol is tended at regular hours to the sound of chank shells and gongs, while offerings of fruit and flowers are presented by the laity (Risley, II, p. 340).

Chanks to be used as wind instruments are chosen of as large size as procurable, often 8 inches long by 4 inches in diameter. The only preparation they require is to have the extreme apex removed, usually by hammering. No tune properly so called can be played, but the tone is capable of much modulation by the lips, and the long drawn notes as they drone clear and mellow on the evening breeze have a haunting charm that clings sweet and seductive in the memory ; it has a mystic wail perfect in appropriateness to its religious use before the shrines of the gods of a profoundly philosophical creed.

Sinistral shells whenever possessed by a temple, are usually mounted in handsomely decorated golden settings and used as libation vessels in the service of the god. Whether the god be Siva in the form of a lingam, or Vishnu or other deity represented in anthropomorphic shape, the officiating priests must lave it with water rendered sacred by being poured from the mouth of a chank.[2] On certain auspicious days cow's milk is used for libations in lieu of water. And if the doubly sacred sinistral chank is not possessed by the temple, then a choice example of the ordinary form must be used.

In family devotions the chank is also employed as a libation vessel by strictly devout Brahmans, both Saivites and Vaishnavites. Daily before the mid-day meal

---

[1] Day, "Music and Musical Instruments of S. India and the Deccan," 1891.

[2] A Tamil proverb says : "If you pour water into a chank it becomes holy water ; if you pour it into a pot, it remains merely water."

the Brahman head of the family, after taking his bath, prostrates himself before the family shrine and then, chanting some hymns from the Vedas, pours water over the image of the deity from the mouth of a chank shell. Then he dresses the god and commences his prayers.

In Thibet the call of the chank is amongst the most familiar sounds to be heard in the monasteries and temples of the Lamaistic faith. The writings of travellers in that most priest-ridden of countries, contain frequent mention of the custom. Sven Hedin for example, when describing the opening ceremonies of the Losar or new year festival which he saw in the great monastery of Tashilunpo in Shigatze—the seat of the Tashi Lama, says :—" Suddenly from the uppermost platforms on the roof ring out deep, long drawn-out blasts of horns over the country ; a couple of monks show themselves against the sky ; they blow on singular sea-shells, producing a penetrating sound, which is echoed back in shrill and yet heavy tones from the fissured rocks behind the convent ; they summon the Gelugpa, the brotherhood of yellow monks, to the festival."

Tea-drinking among the Lamas must never be missed ; the monks partake of it even in the midst of the most important ceremonies, and to prevent the terrible misfortune of a brother being too late for any distribution of tea, the departure of the novices from the kitchen bearing their loads of hot tea in large copper vessels on their shoulders, is signalled to all in the various halls and cells by a loud call upon a chank-horn from the temple roof.

Sven Hedin also describes (" Trans-Himalaya," Vol. II, p. 19) a cave inhabited by a hermit reputed to be one hundred years old, who passed his days crouching in a niche in the wall continually saying his prayers and occasionally blowing a faint blast on a chank.

And when a monk, no longer able to answer the shell's call to gather with his brethren round the teapots and the bowls of tsamba, passes quietly away, the same sound summons those who remain to attend his funeral mass.

In the purer Buddhism of Ceylon the chank cult also finds place, and figures prominently among the musical instruments employed to lend *éclat* to the periodic procession (*perahera*) of the tooth relic at Kandy.

(b)—BRANDING.

All Sri Vaishnavite Brahmans, irrespective of profession, are expected to undergo a ceremony of initiation into Vaishnavism after the Upanayanam ceremony of investment with the sacred thread, in the belief that it is the duty of all of their creed to carry throughout life a memorial of their god upon their person. To effect this, resort is had to branding with heated copper seals made in the conventional form of the various symbols of Vishnu.

Members of this sect are not compelled to undergo this ordeal more than once during their lifetime, but the Madhva sect which comprises chiefly Canarese-speaking Brahmans, have to submit to it as often as they visit their Guru. Men of other castes who become followers of a Vaishnava or a Madhva Achārya (Guru) are expected to present themselves before the Guru for the purpose of being branded. But the ceremony is optional and not compulsory as in the case of a Brahman. Even the women in Vaishnavite families must submit to this branding ; in their case it takes place after marriage in the case of Sri Vaishnavites, while among the Madhvas one form of branding should be performed at any age before marriage should the Guru visit the neighbourhood, and a more formal one again after marriage. Regarding Sri Vaishnavites, Thurston (I, 370) states that " the disciples after a purificatory bath and worship of their gods, proceed to the residence of the Achārya, or to the mutt where they are initiated into their religion, and branded with the chakra on the right shoulder and with the chank on the left. The initiation consists in imparting to the disciple, in a very low tone, the Mūla Mantram, the word Nāmonarā-yanāya, the sacred syllable Ōm, and a few mantrams from the Brahma Rahasyam (Secrets about God). A person who has not been initiated thus is regarded as unfit to take part in the ceremonies which have to be performed by Brahmans. Even close relations, if orthodox, will refuse to take food prepared or touched by the uninitiated."

As Vaishnavite Gurus are few in number, it is necessary for them to peregrinate the country, halting at suitable centres to brand those of their followers living in the neighbourhood just as a Bishop in certain Christian churches tours his diocese to afford confirmation (*i.e.*, initiation) services at periodical intervals. In populous districts where Vaishnavites are in numerical strength the scene at each of the Guru's halting places is intense with interest. Thousands of his disciples gather round eager to be branded. Brahmans are there in force, but men of many other castes and even Paraiyans are there. The ceremonies begin by the making of a fire in a mud pot (*homa kunda*), accompanied by the chant of hymns and the offering of prayer to Vishnu. As Brahmans present themselves for the rite the Guru lifts the copper brands which have been heating meanwhile in the fire and applies them to the shoulders of the people, the chakra on the right and the chank on the left. As each stamp is made the Guru's assistant, usually a Dasari or Vaishnavite mendicant, smears the burnt spots with a paste of *namakkatti*, the same white clay used by Vaishnavites when painting the *namam* or sect mark (improperly called caste mark) on their forehead.

Paraiyans and low caste disciples may not be branded directly by the Guru ; in their case he heats the instruments and hands them to the Dasari, his assistant, who performs the actual operation.

With regard to the branding customs of Madhva Vaishnavites, who believe that to carry a lasting outward and visible sign of their deity on their body helps them to obtain salvation through him, Thurston (I, pp. 371-373) supplies an interesting account :—
" Madhvas have four mutts to which they repair for the branding ceremony, viz., Vaya-

saraya, Sumathendra, and Mulabagal in Mysore, and Uttaraja in South Canara. The followers of the Uttaraja mutt are branded in five places in the case of adult males, and boys after the thread investiture. The situations and emblems selected are the chakra on the right upper arm, right side of the chest, and above the navel ; the chank on the left shoulder and left side of the chest. Women and girls after marriage, are branded with the chakra on the right forearm, and the chank on the left. In the case of widows, the marks are impressed on the shoulders as in the case of males. The disciples of the three other mutts are generally branded with the chakra on the right upper arm, and chank on the left. As the branding is supposed to remove sins committed during the interval, they get it done every time they see their Guru. There is with Madhvas no restriction as to the age at which the ceremony should be performed. Even a new-born babe, after the pollution period of ten days, must receive the mark of the chakra, if the Guru should turn up. Boys before the upanayanam, and girls before marriage are branded with the chakra on the abdomen just above the navel. The copper or brass branding instruments (mudras) are not heated to a very high temperature, but sufficient to singe the skin, and leave a deep black mark in the case of adults, and a light mark in that of young people and babies. In some cases, disciples who are afraid of being hurt, bribe the person who heats the instruments ; but as a rule, the Guru regulates the temperature so as to suit the individual. If, for example, the disciple is a strong well-built man, the instruments are well heated, and, if he is a weakling, they are allowed to cool somewhat before their application. If the operator has to deal with babies, he presses the instrument against a wet rag before applying it to the infant's skin. Some Matathipathis (head priests of the mutt) are, it is said, inclined to be vindictive, and to make a very hot application of the instruments if the disciple has not paid the fee (guru-kānika) to his satisfaction. The fee is not fixed in the case of Sri Vaishnavas, whereas Madhvas are expected to pay from one to three months' income for being branded. Failure to pay is punished with excommunication on some pretext or other. The area of skin branded generally peels off within a week, leaving a pale mark of the mudra, which either disappears in a few months, or persists throughout life. Madhvas should stamp mudras with gopi paste[1] daily on various parts of the body. The names of these mudras are chakra, chank or sankha, gātha (the weapon of war used by Bhima, one of the Pāndavas), padma (lotus), and Narāyana. The chakra is stamped thrice on the abdomen above the navel, twice on the right flank, twice on the right side of the chest above the nipple, twice on the right arm, once on the right temple, once on the left side of the chest, and once on the left arm. The chank is stamped twice on the right side of the chest, in two places on the left arm, and once on the left temple. The gātha is stamped in two places on the right arm, twice on the chest, and in one spot on the forehead. The padma is stamped twice on the left arm, and twice on the left side of the chest. Narāyana is stamped on all places where other mudra marks have been made.

[1] Properly gopi chandiram, a paste made of white kaolin mixed with sandalwood.

Sometimes it is difficult to put on all the marks after the daily morning bath. In such cases, a single mudra mark, containing all the five mudras, is made to suffice. Some regard the chakra mudra as sufficient on occasions of emergency."

So far as I can learn the branding instruments which are employed to sear the two chief symbols, chank and chakra, by means of heat are usually made of copper. In other localities brands of different metals appear to be used as Risley (II, p. 339) states that the Rāmānuja, a Vaishnavite sect in Bengal, when undergoing the initiatory rite (*tapta-mudra*) are branded with the chakra on the right shoulder and the chank on the left, by means of a brand made of eight metals (ashta-dhatū)—gold, silver, copper, brass, tin, lead, iron, and zinc.

Various deviations from the standard ceremonial exist in certain districts; among these is that followed by the Bedar or Boya caste of the Southern Deccan, a caste which largely constituted the old fighting stock of this district. Among them the men are branded on the shoulders by the priest of a Hanuman shrine with the sign of the chank and of the chakra, in the belief that this will enable them to go to swarga (heaven). Female Bedars who are branded become Basavis (temple women) and are dedicated to a male deity and called Gandu Basavis or male Basavis (Thurston, I, p. 194).

This branding of temple girls, or Dēva-dāsi as they are termed in the Tamil country, with symbols of the chank and chakra is always an essential feature in the ceremonies which mark their dedication to the god of their temple, whom thenceforward they serve with dance and song.

Allied to branding is tattooing. The Tandans of Malabar, a caste about the level of the Tiyyans, adopt this method to show devotion to the deity, and among the religious symbols worked into the skin of their arms is that of the chank (Thurston, VII, 10).

(c) THE MENDICANT'S CONCH.

Beggars throughout India occasionally use the chank shell as a musical instrument, and with certain castes of religious mendicants it is an essential part of their professional paraphernalia, so much so that a Tamil proverb likens things in continual association to " the breech of the chank and the mouth of the mendicant."

The Dasari, who belongs to a caste of Vaishnavite mendicants well represented in the Madras Presidency, is often seen in North Arcot and the Southern Deccan, announcing his arrival in a village by blasts on the chank-shell which in that part of the country is one of his five insignia. In Telugu districts the Dasaris are more secular and less religions, and the caste is known as Sanku Dāsari or vulgarly Sanku jadi, the chank-blowing caste.

A mendicant's conch sometimes has the apical orifice mounted in brass; temple conchs are usually without any ornamentation, but the Udipi temple owns one very handsomely mounted in brass and this is sounded whenever the god (Krishna) is carried in procession in the temple car.

(d) DEDICATION OF TEMPLES AND HOUSES.

Wherever a new temple is built, or when a new shrine or god is established and added to the number already there, its dedicatory ceremonies include as one of the most important a special libation to the god from the mouths of 108 chanks, or better still, from 1,008 chanks if so many can be afforded, filled with water and flowers.

The building of an ordinary house in the Tamil country must also have its ceremonial dedication at the time the foundation trenches are dug, though among low caste people the rite consists merely of a superstitious act to ensure good luck or to baulk the evil eye. It is carried out with the help of the sacred chank which thus is seen to touch the lives of the people at every point from the cradle to the grave. Before a single stone in the foundation is laid, the ceremony is carried out on a day carefully chosen as being highly auspicious. A chank is then buried beneath the first stone laid. An old reference to this occurs in a petition quoted by Wheeler (fide Thurston, III, p. 147) from two natives of Madras, in connection with the founding of a village called Chintadrepettah, now a populous division of Madras City. The entry runs :—" Expended towards digging a foundation where chanks were buried with accustomary ceremonies." Roman Catholic converts from low castes follow this custom, as well as Hindus ; in Tuticorin, if a Roman Catholic Parayan desires to build a house, the carpenter employed by him chooses an auspicious day by reference to a native calendar, a chank is bought in the bazaar and on the day chosen, having dug a foundation trench and prepared at the bottom a bed of coral stone and mortar, the chank is laid thereon. In the cavity of the shell small fragments of five metals (panjalokam), gold, silver, copper, iron and lead, are placed, turmeric and sandalwood water is sprinkled over, and then the whole is hidden under a mass of sweet-smelling flowers. The ceremony is ended ; the first stone may now be lowered into place upon the chank and its contents, and good luck is believed to be assured to him who will inhabit the house.

It may, however, happen that in spite of every precaution, an inauspicious site appears to have been chosen as shown by a sequence of misfortunes happening to the householder. In such cases Hindus may perform a special ceremony called Chankusthabanam to remedy the evil. A chank-shell is filled with water and incantations made for forty-five days. At the end of this period of propitiation, the chank is buried under the house wall. (Winslow, " Tamil and English Dictionary," p. 390, Madras 1862.)

Among the Parawa caste living in Tuticorin and other coast towns and villages in Tinnevelly and Rāmnād, misfortune is often sought to be averted from the individual by almost completely burying a chank shell in the floor of the hall (kūdam), about two feet on the inner side of the doorway to the street. A small portion of the back of the shell shows as a patch of white on the surface of the floor. The explanation of the custom current among Parawas is that the shell is so placed that when an inmate leaves the house, he must pass over it, usually touching it, and this will prevent misfortune happening to him.

(*c*) MARRIAGE CEREMONIES.

The chank has important but variable functions to perform at weddings among all Hindu non-Brahman castes in the districts of the south of India, where this shell is blown by the barber (ambattan) particularly at or immediately after the tying of the *tali* or marriage badge around the bride's neck ; the bridal couple usually occupy a raised platform, and round and round this the barber walks while blowing his chank.   In Bengal this custom of chank-blowing during weddings is even more general ; a common formula which runs

> " Ganga ka pani samundra ki sank
> Bar kanya jag jag anand."
>
> (" May Ganges water and sea-chank betide
> Enduring bliss to bridegroom and bride.")

is usually recited during the marriage (Risley, II, p. 116) with reference to the blasts on the conch which accompany the ceremonies—the equivalent of the marriage bells in Christian ritual.

In Telugu districts the chank is not used by any caste, non-Brahman as well as Brahman, during weddings, as this is considered inauspicious because chank-blowing is specially associated with funeral ceremonies.

Usually a man of a special caste is engaged to blow the chank at the customary times ; in the Tamil country the caste barbers (ambattans) perform this duty ; among the Telugus the chank blower is usually a Dasari, among the Uriyas, a Ravulo.

Sometimes, however, women of the family or of the caste perform the chank-blowing duty.   Among Bengal Brahmans, for instance, one section of the elaborate and lengthy marriage ceremonies consists of a procession of seven married ladies headed by the bride's mother going round the bridegroom seven times, some sprinkling libations of water and vociferating the hymeneal cry of *ulu-ulu*.   One of the seven carries a conch and blows it as she goes (Risley, I, p. 150).   A custom somewhat akin to the above prevails among the Kallan caste of Tanjore, Madura, Trichinopoly and Tinnevelly.   On the wedding day the sister of the bridegroom goes to the house of the bride, accompanied by women, some of whom carry flowers, coconuts, paddy, turmeric, milk, ghee, etc.   On the way two of these women blow chank shells. (Thurston, III, p. 80.)

In passing it is interesting to note that a section of this caste, the Puramalai nādu Kallans, practise the rite of circumcision probably as a survival of a forgotten forcible conversion to Muhammadanism.   The rite is carried out in a grove or plain outside the village and *en route* to the place, and throughout the ceremony a chank is blown at frequent intervals. (Thurston, III, p. 74.)

It is noteworthy that Brahmans in the Tamil and Telugu districts do not employ the chank during marriage ceremonies though their brethren in Bengal do. Among Telugu Brahmans living in Uriya districts the custom of Bengal used to be followed at marriages, but this is gradually dying out ; as one Brahman in Berhampur (Ganjām) remarked " The present day Brahmans here have more regard for the magic flute than for the divine voice issuing from the chank."

In Bengal the association of the chank with marriage is more intimate and deep than elsewhere ; no Bengali lady is properly or legally married unless chank bangles, which should be lacquered red, be placed upon her wrists. In the Madras Presidency, marriage bangles are used only by a few sections of the agricultural and pastoral castes (Vellalans and Idaiyans).

One of the most interesting facts brought to light during the present research is the weighty evidence we have that in former days the tali, the essential marriage symbol among Tamils, was directly connected with the chank, either composed of a piece of the shell or of a metal ornament in the form of a miniature chank shell. We find this marriage badge named specifically *sankhu tali* among four castes widely separated both geographically and in status and civilisation. First are the *Chanku tali Vellalans*, a section of the great Vellalar caste, who wear, according to Winslow (" Tamil and English Dictionary," Madras, 1862), a representation of the chank on either side of a central symbol. Unfortunately, apart from this reference, I have been unable to trace the location of these Vellalans at the present day, or to obtain any details of the custom.

Two other castes with the same marriage badge occur on the West Coast, and it is significant that one is undoubtedly of Tamil origin. This is an immigrant branch of the Idaiyans known locally as Puvandans, settled in Travancore. On ceremonial occasions the women wear the Tamil Idaiyan dress while in ordinary life they attire themselves after the fashion of Nayar women. Their tali is known as *sankhu tali* and a small ornament in the form of a chank is its most conspicuous feature. (Thurston, II, 366.)

The other West Coast caste using a *sankhu tali* is that of the Thandan Pulayans, a small division of the Pulayans who dwell in South Malabar and Cochin. The women dress in a leaf skirt made from the stems of a sedge called *thanda*, which are cut into equal lengths, woven at one end and tied round the waist so that they hang down below the knees.

According to Ananthakrishna Aiyar (Thurston, VII, p. 23.) : " At the marriage ceremony, the tāli (marriage badge) is made of a piece of a conch shell (*Turbinella rapa*) which is tied on the bride's neck at an auspicious hour. She is taken before her landlord, who gives her some paddy, and all the coconuts on the tree beneath which she happens to kneel. . . . . . . To ascertain whether a marriage will be a happy one, a conch shell is spun round. If it falls to the north, it predicts good fortune ; if to the east or west, the omens are favourable ; if to the south, very unfavourable."

Lastly, and most interesting of all, we find a caste calling their marriage badge *sankhu tali*, which on examination shows no likeness to a chank shell. These are the Parawas of the coast towns on the Indian side of the Gulf of Mannar. When the Portuguese arrived there early in the sixteenth century, these people, who were principally pearl fishers, chank divers, and fishermen, were orthodox Hindus, but the stress of Muhammadan competition drove them into alliance with the Portuguese and they went over in a body to the Roman Catholic church. To-day the badge tied around the bride's neck on marriage consists of three ornaments, a central cross flanked on either side by the symbol of the Holy Ghost ; nevertheless it is called *sankhu tali* as among the castes first mentioned. There is no doubt that when the caste was a Hindu one the *tali* was true to name, indeed Parawa tradition is definite, for it asserts that originally the central ornament was a small figure of some Hindu God (probably Krishna) flanked by one of a chank shell on each side. The use of the original name is a strange persistence in view of nearly 400 years' sojourn within the Christian fold; it is one of the many signs of tolerance shown by the Roman Catholic priesthood towards their converts' prejudices on immaterial points—a tolerance in petty matters that has done much to help that church in its propaganda.

Among some castes, including the Bauris and Dandasis of Ganjām, turmeric water from a chank shell is poured seven times over the hands of bride and bridegroom, which are tied together with seven turns of a turmeric-dyed thread. (Thurston, Vols. I. and II.)

(*f*) DEATH CEREMONIES.

Throughout the Tamil country all non-Brahman castes which observe Hindu rites have the chank sounded as the body is being taken to the funeral pyre or to the burial ground. It is usual also to employ the conch-blower on the last day of the sraddh ceremonies in those castes which follow the orthodox ritual. Among the Telugus these same rites are largely followed, but it is said that Vaishnavites do not observe them. Among both races, the Brahmans do not have the conch blown at any period of the obsequies—a sign that lends weight to the theory that the chank has been borrowed by Brahmanism from another religion.   •

In the Madura and Tinnevelly districts the conch-blowers at a funeral are Ambattans or barbers, the same caste as performs likewise at weddings. Among the Idaiyans of these and the neighbouring districts one part of the funeral rites consists in the son perambulating the pyre thrice with a pot of water on his shoulder ; at each turn the barber makes a hole in it with a shell when the head of the corpse is reached. Finally the pot is broken near the head. (Thurston, II, p. 362.)

Further north, in the East Coast districts from Tanjore and Salem to the Kistna River, the Panisavans are by caste custom the funeral conch-blowers ; they may indeed be accounted the undertaker caste, as it is their duty to carry news of the death to the relations of the deceased. It is they who generally keep all the materials necessary for the funeral including the palanquin required for the conveyance of the corpse to the cremation ground. At the funeral, the Panisavan follows the corpse, blowing his conch. When the son goes round the corpse with a pot of water, the Panisavan accompanies him sounding his conch the while. On the last day of the death ceremonies (Karmand-hiram) the Pansivan should also be present and blow his conch especially when the tali is removed from the widow's neck. (Thurston, VI, p. 56.)

The insignia of the Panisavans are the chank and the tharai, a long straight trumpet.

In Coimbatore district, the duty of sounding the death conch belongs to the members of an important sub-division of Paraiyans, called on this account, Sankhu Paraiyan. (Thurston, VI, p. 81.) In Travancore when a headman or kaikkaran of the Paraiyans settled there happens to die, a chank-shell is buried with the corpse. (Thurston, VI, p. 134.)

The chank sometimes has a place in the death ceremonies of castes which are not Hinduised. Thus among the Cherumans of Malabar and Cochin, a caste of agricultural serfs, according to Mr. Ananthakrishna Aiyar (Thurston, II, p. 81.), " The son or nephew is the chief mourner, who erects a mound of earth on the south side of the hut, and uses it as a place of worship. For seven days, both morning and evening, he prostrates himself before it, and sprinkles the water of a tender coconut on it. On the eighth day, his relatives, friends, the Vallon, and the devil-driver assemble together. The devil-driver turns round and blows his conch, and finds out the position of the ghost, whether it has taken up its abode in the mound, or is kept under restraint by some deity. Should the latter be the case, the ceremony of deliverance has to be performed, after which the spirit is set up as a household deity."

How far the conch is used in funeral rites outside the Madras Presidency, I am not in a position to say, except in regard to Thibet, where as already incidentally mentioned, it is a custom to sound it as the body of a monk or a nun is being conveyed from the place where death occurred.

(g) TOTEMS.

Totems as the distinctive signs of exogamous septs must have been at one time universal among tribes of Dravidian origin. To-day a well developed totemistic system characterises the tribal organisation of the Santals and Oraons who retain languages distinct from those of the surrounding peoples, know nothing of the caste system, and who continue to worship non-Aryan gods. Among the Santals 91 septs are known, and one of these is known as Sankh. The members of this sept may not cut, burn, nor use the shell, nor may the women of this sept wear it in personal adornment.

Above these still primitive tribes and between them and the fully Hinduised peoples who are split up into castes, are a large number of partially Hinduised tribes which in many cases show distinct traces of a totemistic organisation. Among these the

c

Kurmis of Bengal, and the nomad Koravas who wander throughout the peninsular part of India, both have an exogamous sept or gotra of which the totem is the chank-shell. Among the Kurmis this sept is called Sankhawar ; its members are prohibited from wearing ornaments made from chank-shells. With the Koravas it is termed Samudrāla, signifying the sea, and people of this sept may not use the chank in any way. Higher than these are the Kalinjis, an Uriya agricultural caste, and the Kurubas, a caste of shepherds and weavers widely spread throughout the Madras Presidency. Both castes comprise septs named after the chank, in the case of the Kalinjis, Sankho, in that of the Kurubas, Sankhu. I am not aware whether the septs among the former caste have now totemistic value, or if it has become merely a name, a gotra name ; in any case it may be taken as certain that in the pre-Hinduised condition, the name of the gotra was of real totemistic value. Bhāgo (tiger) and nāgo (cobra) are names of two other gotras of obviously totemistic origin. With the Kurubas, the sept is undoubtedly exogamous and its totemistic character certain.

Another caste or sub-caste showing by the names of its sections a probable totemistic origin is that of the Koppala or Toththala, a sub-division of the Velamas, a caste of agriculturists in the Vizagapatam district. Among their sections are some named Nāga (cobra), Sankha (chank), Tulasi (basil or tulsi) and Tāblēu (tortoise). At the present day these divisions although apparently of totemistic origin, have no significance so far as marriage is concerned. (Thurston, VII, p. 340.)

(h) EVIL-EYE SUPERSTITIONS.

Belief in the reality of the malign results which ensue from being overlooked by the evil-eye is frequently present in an acute form in the Madras Presidency. It is specially dreaded in the case of houses under construction and in respect to valued cattle. Everywhere in Tamil districts the custom prevails more or less extensively of seeking protection for draft bullocks by tying a small chank-shell upon the forehead of such as are in good condition or in any way specially valuable or beautiful in their owner's eyes. Of late years the custom has tended to fall into abeyance in certain districts. As is to be expected the people of country villages cling to it with greater tenacity than those in towns. Many, however, decorate their bulls in this way without thought of it as an amulet against evil—to them it is merely an old custom to be followed, or else they put it on their favourite animals as an ornament to mark the pride they have in them. Again, some, from a peculiar shyness often met when discussing such matters with the peasantry, deny that the chank-shell is used as an amulet although in reality it may be so used by them.

In the southern Deccan the custom appears to be falling more quickly into abeyance than in Tamil districts. The Collector of Kurnul informs me that though the practice survives in parts of Dhone, Kumbum, Koilkuntla and Sirvel taluks of tying a chank on the forehead or round the necks of bullocks and ponies, it is gradually dying out ; in Bellary the Collector states that it still prevails in Bellary, Hospet and Hadagalli taluks in respect to bullocks, but adds that the people do not now attach any religious or secular significance to it. In Anantapur the Collector states that the custom is not now followed in that district.

In Madras City it is quite common to see it and there also I have seen a shell hung by a chain or a cord round the neck of a cow-buffalo when in milk to prevent her being " overlooked " and her milk thereby dried up prematurely. In country villages this latter custom is not infrequent both in regard to ordinary cows and to cow-buffaloes. In the Madura, Trichinopoly, Salem and adjacent districts a shell is often hung round the necks of jutka and pack ponies, not only by Hindu owners, but also by Muhammadans. In Madura specially valued sheep occasionally are similarly guarded from evil, and in the same district I have seen milch-goats protected in a like manner. In all these cases the shells used are of small size, the great majority being dead or sub-fossil shells from the mud-beds of Ceylon. They are the same as are sold as feeding spouts for infants in every Tamil bazaar. A hole is bored or broken in the back and the rope passed through this and out at the mouth of the shell. The surface is generally roughly engraved in a coarse spiral or scroll pattern.

Used probably for a similar purpose may have been the handsomely engraved large chank shell obtained from an oblong sarcophagus of red pottery found in the prehistoric burial site at Perambair in the south of Chingleput district. The size is much greater than any I have ever seen used to decorate ordinary cattle ; no ordinary person owned it we may be certain—probably it decorated the forehead of a bull or possibly of an elephant belonging to a man of great local importance. From the same tombs came three other handsomely decorated chank ornaments, two of which were probably ornaments for the hair (see p. 41 for further particulars).

Once only have I seen [1] more than a single shell hung round the neck of any animal, but that in ancient times a different habit at least occasionally prevailed is possible, to judge from the string of 16 small chank shells from a barrow near Guntakal junction in the Anantapur district, now to be seen in Madras Museum. All these shells have had the apices broken in and partly rubbed down and each has the thickest part of the body perforated from side to side so permitting them to be strung together. They appear to have formed a necklace, but whether they were suspended round the neck of a bull or may be hung like a chain of office round the neck of some person of importance, we have no means of determining.

In Malabar the chank is little in evidence, but Logan ("Malabar Manual," 1887, Vol. I, p. 175) records a belief there prevalent that a cow will stop giving milk unless a shell (not necessarily a chank) is tied conspicuously about her horns and at Tanur, Malabar, I have seen valuable sheep with shells other than chanks hung round the neck.

[1] At Tirupalakudi, on Palk Strait.

Further north, in the coast villages of South Canara, south of Mangalore, rings made from Strombus shells but known locally as chank rings are employed by parents to avert the evil eye from their young children. At Kasargod, Bekal and the adjacent villages I have found the custom especially common among the Mukuvans, a caste of immigrant Malayali fishermen. Children from 3 to 4 years old in these villages are frequently given necklaces made of Strombus rings alternating with elongated glass beads. Some other castes in the same neighbourhood, Mokayans and Tiyyans together with some Mappillas, are said to follow the same custom. Usually the rings do not exceed twelve in each necklace. Adults do not wear these amulets as is the habit of the woman of certain sections of the Cheruman caste in Malabar (cf. p. 39) and of the Hill Vedans of Travancore. How far these and other facts connote former wide or even universal prevalence of this habit among the indigenous population of Malabar, is a line of inquiry likely to repay careful investigation.

Finger rings purporting to be made from chank shells, but usually cut from a small species of *Strombus* common on the western coast of the Gulf of Mannar, are also used very freely throughout the Tamil country and also in Malabar and Cochin, chiefly by non-Brahmans among Hindus, as amulets against evil spirits, the evil eye and certain sicknesses. In Tinnevelly, Madura and Rāmnād the custom is very prevalent among both sexes of non-Brahmans. Labbai and Marakkayar Muhammadans in whose veins much Hindu blood is present, also affect the custom. The Vellalans, although like Brahman adults, they wear, except in one section, neither chank-bangles nor rings, often provide their children with chank-rings or else with pieces of chank-shell tied on the wrist of the right hand by means of black thread, as an amulet against the disease called chedi which is, I believe, rickets. In some cases the ring or piece of chank is placed on the wrist only when the disease has laid hold of the child, in others it is tied on when the child attains its second month and kept there till it is three years old, when it is believed that all danger of contracting the disease has passed. Among the castes ranking next—Chettis, Kollans, Thachchans, Thattans, Naidus, Idaiyans, and Chaluppans, a chank-ring is often worn as an amulet against pimples on the face ; occasionally their young children are provided with small and roughly ornamented chank-bangles to safeguard them against *chedi*. The low castes or Panchamas such as Pallans, Vallayans, Paraiyans, etc., are the most regular devotees of these amulets.

The Roman Catholic Parawas of Tuticorin and the other Parawa strongholds on the Pescaria Coast have also been great believers in the virtue of chank amulets, and till recently all babies were given chank-bangles to protect from convulsions and from *chedi*. Even now the poorer and more ignorant continue to employ these amulets, keeping them on the wrists for about three years. The richer and better educated have either abandoned the practice or keep the bangles on for a much abbreviated period. The Parawas formerly also employed pieces of the curious egg-capsule of the chank for the same purpose as the bangle, a fragment of the capsule (*chanku-pu*, literally " chank-flower ") being tied by means of thread upon babies' wrists.

In Madura chank amulets are used even more freely than in Tinnevelly. In addition to bangles and rings used as amulets against the evil eye, or ailments such as *chedi* and pimples, very roughly fashioned and imperfectly rounded fragments of chank shells are used in the manner of beads to make necklaces which are used as amulets. Maravans, Paraiyans, and Chakkiliyans are among the castes chiefly addicted to the wearing of these and other chank amulets ; these people often give their children both chank-bangles and necklaces of chank-beads with a view to multiplying the counter-vailing influences against the evil eye and against disease.

Chank-bead necklaces (*chanku malai*) are also worn largely by children of the poorer Chettis and of the Vanniyans (oilmongers) who, though they do not generally wear chank-bangles, will wear these chank-bead necklaces. The people of the lower castes also use the same rough beads to make bracelets (*chanku pasi*) worn on the wrists for the same object as the bead necklaces. Similar customs in regard to rings and bead necklaces prevail in Tanjore and in South Arcot, where the low castes, especially Vanniyans, Koravans, Paraiyans, Chakkiliyans and wandering Lambadis, generally wear them as amulets against evil spirits, the evil eye and sickness. Koratti women appear to be the only ones in South Arcot who wear bead bracelets in addition to other bangles according to information kindly supplied by the Collector, Mr. Azizuddin.

The trade in these amulets is of considerable dimensions. Thus the Tahsildar of Chidambaram reports that rings worth about Rs. 500 are sold yearly in his taluk to the residents and to the thousands of devotees who flock to the great temple of Nataraja. The price runs from two pies to one anna per ring—some are very rough untrimmed *Strombus* rings, while the higher priced may be of real chank with a few oblique lines of ornament sawed or filed on the outer surface.

The rough beads used in making necklaces and bracelets look very much like tiny carpal bones ; they sell about eight beads for one pie—a whole necklace may be bought for one anna ; the small bangles worn by babies in Madura and Tinnevelly cost 3 to 5 annas per pair. The latter are made largely in Kilakarai, a large Muhammadan settlement on the Rāmnād coast, near the head of the Gulf of Mannar. The majority are cut chiefly from under-sized shells too small to have any value in the Bengal market. By rights these shells should not be fished ; they should be put back alive by the divers in order to grow to adult dimensions. This precaution is followed as far as circumstances permit in the Tuticorin chank fishery which is conducted departmentally by the Government of Madras ; in the Rāmnād fishery the short-sighted greed of the renter and his employees takes no account of any such precaution for the future prosperity of the fishery.

(i) For personal adornment.

In personal adornment, and apart from any uses it is put to in the form of amulets, the chank-shell is employed principally as the crude material wherefrom beautiful bracelets of many patterns are made for use in Bengal and the adjoining provinces; subsidiary uses to which it is put are to fashion from it finger rings, necklaces, disc ornaments for headdresses and caps, and as a recent addition, coat and dress buttons. The bangle industry in all phases is treated separately and at full length in Section II, to which the reader is referred for details.

Rings actually made from chank-shell are not manufactured in any quantity; their place is taken largely ·by those made from a much smaller shell, a species of *Strombus*, found on the Rāmnād coast of the Gulf of Mannar. So far as I can learn, the industry is localised at Kilakarai, a seaport of Rāmnād inhabited largely by Muhammadan (Labbai) fishermen, pearl-fishers and chank-divers. The rings are usually exported with a minimum of finish; only the roughness of the edge is rubbed off and nearly always the chestnut stippling is clearly recognisable. Many are sold in the bazaars or by peddlers throughout the Tamil and Malayalam districts, usually as amulets against the evil eye and against such minor ailments as pimples on the face and various· skin troubles. So far as I can ascertain, the only people who use these rings in personal adornment are two tribes of low civilisation living in the Malayalam country—the Hill Vedans of Travancore and certain sections of the Cheruman tribe in Cochin and Malabar. The former I have not seen. They are described by Thurston (VI, p. 333) as living in wretched huts and employed chiefly as rice-field watchmen. He states that both the men and women of this tribe wear numerous bead necklaces interstrung or otherwise associated with a few of these rings. In a photograph given by the same authority (Vol. III, p. 177), a man is shown wearing numerous strings of glass beads passed *through* eight Strombus rings. In the case of the women these necklaces hang down as far as the abdomen.

The Cherumans were formerly the agrestic serfs of Malabar, Cochin and Travancore —the Malayalam country or Kerala. To-day they still remain largely in a servile condition, carrying on for their masters the heavy labour of the fields; they receive their pay almost always in kind. They are divided into a considerable number of endogamous sections differing in appellation in different districts—a sign of long-continued residence in a country of difficult intercommunication.

All Cheruman women are greatly addicted to the use of necklaces, particularly of the showy strings of beads now put within the reach of the poorest by the enterprise of Austrian and Italian manufacturers. Of other clothing they wear the scantiest— a very dirty, once white cloth, pendant from the waist, being their usual garb. Certain sections wear as a distinctive badge, in addition to numerous bead necklaces, a long cord whereon are strung large numbers of Strombus rings (*chanku modiram*)—they believe

[Photo by J. Hornell.

Fig. 1.—Group of Cheruman women wearing necklaces of so-called chank-rings.

[Photo by Vividha Kala Mandir.

Fig. 2.—Selling chank shells to pilgrims returning from Bét.

them to be of chank-shell. The bead necklaces are usually wound many times round the neck itself, roughly forming a collar often reaching as high as the chin. The chank necklace is worn at a lower level, and lies on the shoulders and on the upper part of the breast; it looks much like a chain of office and is indeed the badge of the tribal sept. At Tanur (Malabar) where after much trouble three Cheruman women were got together for my inspection, one of the husbands had to be paid a day's wages to keep guard over them to prevent their flight. They were all exceedingly shy, and it was with much reluctance that they stood up in front of my camera. As will be seen by reference to Pl. III, Fig. 1, the chank ring necklaces (chanku modira mala) are made up of a very large number of rings not strung but tied by the upper edge to a strong cord in such a way that each ring overlaps its neighbour on one side and is similarly. overlapped on the other side by the succeeding ring, much as the rings in chain armour are arranged. From 50 to 100 rings are required to form a full necklace of this pattern ; as each ring costs from 3 to 6 pies in the local bazaar, the total cost may amount to Re. 1-8-0 or Rs. 2, a large sum to these exceedingly poor people. The Cherumans who wear these chank chains in the Tanur neighbourhood say they belong to a sept named Kalládi Cherumans and that they wear them to distinguish themselves from the Paliya and other septs with which they may not intermarry. In Tanur bazaar I saw a single example of another pattern of this strange necklace worn by a woman also said to be a Kallādi Cheruman. In this case the number of rings used were comparatively few, 20 in all, and between each pair were strung a couple of glass beads of different colours. Each ring was separated by an interval of about an inch from its neighbour on either side, and instead of being fastened to the common cord by a single loop, it was fastened by two separate loops which enabled it to lie flat upon the skin. The woman shrank against the wall, averting her face and trying to sidle away, and it was with great difficulty she could be persuaded to answer a few particulars. Among other information she gave, was the statement that this necklet is believed to protect from evil spirits.

So far as I have been able to ascertain, these chank necklaces are assumed soon after a girl attains puberty if her parents can afford it. If they be very poor and cannot afford it, then, when her marriage is arranged, it is generally settled that the bridegroom shall provide the needful ornament. There is no special ceremony followed at the time a girl puts on her chank necklet for the first time. As a rule the men of the family attach the rings to the cord.

This custom seems to be losing ground quickly, for while many people knew of it further south in Malabar, I never saw this ornament in use in North Malabar. Many Cherumans were seen between Cannanore and Mount Dilly, but all said few use it now, preferring glass or imitation coral beads for their necklets. There is no doubt that formerly the custom was widely spread among the servile population of Kerala, and as these people's religious beliefs consist almost solely of the dread of malignant spirits, it is extremely probable that originally the necklet was used as an amulet against demons and the evil eye, though now it is more generally considered as a sept badge. The custom of long-settled Malayali immigrants (Mukuvans, etc.) on the South Canara coast, of putting similar necklaces round their children's necks already referred to on p. 36, appears to furnish strong corroboration of this conclusion.

In Bengal a few ornamental finger rings are now made, carved in simple patterns and highly polished. These are not in great demand, and I am uncertain as to whether they are worn as ornaments or as amulets. At Kilakarai a few roughly decorated thin finger-rings to be used as amulets are also produced, in addition to the roughly made, thick and clumsy sections cut from Strombus shells.

The first mention of the use of discs cut from chank-shells to ornament caps and headdresses occurs in Tavernier's " Indian Travels." In 1666 he was in Dacca and records the fact that Bhutan merchants took home quantities of "round and square pieces (of shell) of the size of our 15 sol coins." He also states that " all the people of the north, men, women, girls and boys, suspend small pieces of shell both round and square from their hair and ears."

Whether the trade is as large as in former days, I cannot say. It is now of small monetary value.

The Nagas of Assam, lately brought to prominent notice through the good work they did as carriers during the Abor punitive expedition (1912), employ these discs both to form necklaces and to decorate the handsome plaited cane helmets worn by the men. These latter are conical in shape, about a foot high, and covered with a layer of fur and hair, black or red in colour. When decorated with chank-shell discs, these are arranged as coronals, adding most effectively to the general design (W. Crooke, " Natives of Northern India," p. 47, London, 1907). As the Nagas are known to have set much greater store by the chank in former times, say prior to the middle of the nineteenth century, it is probable that then the use of chank discs as items of ornament was much more general among this race than it is now. Still the custom is quite common, for Mr. Stanley Kemp, who accompanied the Abor expedition as naturalist, informs me that the Naga coolies employed as carriers frequently wore necklaces formed of square concave portions of chank-shell with a large cornelian set en cabochon in the centre. Sometimes long cylindrical beads made from chank shell, tapered slightly at either end, were used instead and cornelian beads were often seen in conjunction.

In the middle of last century Major John Butler mentions (" Travels and Adventures in the Province of Assam," p. 148, London, 1855) that at sixteen years of age a Naga youth " puts on ivory armlets or else wooden or red-coloured cane ones 'round his neck. He suspends conch shells with a black thread " (round his neck) " puts brass ornaments into 'his ears and wears the black kilt ; and if a man has killed another in war he wears three 'or four rows of cowries round the kilt." From a specimen of chank-shell necklace from the Naga hills contained in the ethnological collection of the Indian Museum, Calcutta, it appears that the shells before being used were bisected longitudinally, each half being hung as a pendant by one end from the cord encircling the neck, the whole forming a most uncomfortable-looking decoration, particularly as the custom is to wear them slung at the back of the neck.[1]

Sixty years ago chanks constituted the currency of the Naga tribes, but with the advent of the rupee, the consideration in which these shells was held largely disappeared, and now these quaint chank necklaces are seldom worn. Mr. Kemp saw them worn on only one or two occasions during the Abor expedition (1912).

At death these ornaments and all the other items of the deceased's dress, together with all his treasured weapons, are laid upon the grave.

Among the Abors the custom of wearing chank ornaments must be very rare, for Mr. Kemp, who most kindly gave attention to this subject, saw only a single instance—a Gam or headman of Komsing village, who was found wearing a necklace composed of round concave discs of shell.

The furthest point east to which I have been able to trace the use of chank discs is the banks of the Upper Mekong to the northward of Tali-fu in the Chinese Province of Yunnan. Here Prince Henri d'Orléans (" From Tonkin to India," p. 174, London, 1898) found the women of the wild Lissu tribe, a branch of the Lolo race, " often naked to the waist ; they had a little hempen skirt and a Chinese cap decked with cowries and round white discs which are said to be brought from Thibet and looked to me as if cut out of large shells." In some villages they wore a heavy turban in place of the little white-disc'd cap.

The finest discs I have seen are prehistoric in age, having been taken from the very peculiar oblong sarcophagi, made of red pottery and raised on 6 or 8 stumpy legs, from the ancient graves at Perambair in the south of Chingleput District, near Madras. These discs, two in number, now on exhibit in the Madras Museum, are respectively about 2½ and 3 inches in diameter ; in both there is a small central perforation. They appear to have been cut from the belly of large shells as the convexity is not great. The convex surface in each case is ornamented with geometrical patterns (different in each case) of much delicacy. One is illustrated on Pl. XXXIII. of the report for 1908-09 of the Director-General of Archæology.

The shape and size of these ornaments and the character of the incised patterns suggest that they have been used as boss ornaments for the back hair in the manner affected by native women of certain castes in South India. Until I had seen these ancient chank ornaments I had never heard of the chank shell being used for this purpose, but subsequently I have been told that the custom still survives in Travancore and that

[1] Similarly bisected chanks hung by a cord round the neck are also seen among the Chins of the Central and Northern sections of the Chin hills in Burma. My informant, Mr. W. Street, of the Burma Commission, states that the women alone wear this neck ornament ; usually a single shell is used and apparently fresh supplies no longer come into the country as those now worn are heirlooms in the families of the wearers. It is probable that cessation of the supply synchronised with the discontinuance of chank shell currency among the Naga tribes living to the north of the Chin country.

when the wearers cannot afford gold these boss ornaments for the back hair may also be made from ivory, bone, horn and even coconut-shell. These are usually richly carved and frequently mounted in gold. The central hole in the chank disc would in such cases be used to secure the ornament to the hair.

Beads made from chank-shells do not seem to be used except to form bracelet and necklace amulets. I have seen no carefully worked and polished beads suitable for purely ornamental use. It is possible, however, that necklaces have been made from the pearls which are occasionally, but very rarely, found in the flesh (mantle) of the chank. Such pearls are not uncommon in the West Indian conch which produces them in sufficient frequency to constitute them regular items in the jewellery trade. These "pink pearls" as they are called, are usually made up into necklaces. The Indian chank is a much smaller shell, and although fished in far greater numbers than the West Indian shell, it is exceedingly rare for a pearl to be found. The colour of the few found varies from porcelain white to pale pink, and while it would be a matter of the greatest difficulty to obtain enough during many years' search to make a necklace, matching the colour and grading the size of the pearls to make the ornament a thing of beauty, is well-nigh an impossibility. I have three of these chank pearls in my possession ; they are the only ones I have ever seen. The largest is a perfect sphere, $\frac{13}{32}$ inch in diameter, porcelain or opal white in colour, of lovely skin grained with a most peculiar mottling something after the fashion of the "watering" of watered silk. Another is slightly elongated in one axis ($\frac{5}{16}$ inch × $\frac{1}{4}$ inch), oval or elliptical in outline, of a very pale pink tint and possessing also the peculiar watered grain shown by its fellow. The third is salmon coloured, almost spherical, with a diameter of $\frac{1}{3}\frac{1}{2}$ inch.

A few coat buttons are now made from chank-shells at Dacca—a recent departure on the part of one or two cutters who have made a feeble and ill-sustained attempt to open up new sources of demand. The main obstacle to the success of this new departure lies primarily in the lack of power-tools to cut up, drill and polish the material more cheaply than is possible so long as dependence is placed upon hand labour, however low be the wages paid. Granted even this change, great difficulties in the way of success exist in the lack of artistic versatility characterising the chank cutters' trade and the inability of the ordinary Indian manufacturer to appreciate the value of a judicious advertisement of his wares. He grudges to pay out money in advertisement and when he does so he usually brings about the loss he fears by lack of foresight in keeping up his stock of the advertised article or by the foolish as well as dishonest trick of sending an inferior article to that ordered and paid for by the customer who answers his advertisement.

With power machinery utilised by firms trading on sound and honest principles, there should be a very great field for the sale of chank buttons. There is nearly always a good demand for handsome buttons suitable for the decoration of ladies' jackets and coats and owing to the beautiful porcellaneous appearance of chank-shell when cut and engraved with some attractive or distinctive pattern, suitably designed buttons should meet with appreciation in the European and American dress trade. Rough cut buttons priced at what seems to the European ridiculously high rates are worse than useless, and beyond this the imagination of the Dacca manufacturer cannot soar —at present.

(j) FEEDING SPOUTS.

In the ordinary everyday life of the people of Southern India, the chank subserves several useful functions. Some of these have already been touched upon, but the most useful remains to be mentioned—that of small shells used as feeding spouts when weaning infants. The bazaars in every big Tamil town furnish these primitive utensils, made from undersized shells usually of the sub-fossil description obtained from the muddy lagoons near Jaffna in Ceylon.

The shells are prepared for market by breaking down parts of the inner portion of the terminal whorls just inside the mouth and by removing the central part of the columella. The canal-shaped canaliculum of the mouth is deepened and straightened to form a rough spout ; the exterior surface of the shell is rubbed down and upon it is engraved a rude pattern, usually in the form of a spiral scroll with a few star-shaped emblems ; last of all it receives a thin coating of fine lime or whitewash to hide imperfections and improve the colour (Pl. VI, Fig. 4). For the purpose intended it is quite effective, but how far the crevices of the interior, by offering obstacles to efficient cleansing, harbour and promote the rapid growth of bacteria and so lead directly to infantile diarrhœa, it is difficult to say. If the shell be boiled daily, a very simple precaution and easier to do in the case of a chank than in that of a glass bottle, there would be no danger, but I fear this is seldom thought of. In feeding baby monkeys just taken from their mother I have found this feeding shell most useful ; the sight of the little creature hanging on with both fore-paws to the shell, half choking in its eagerness to swallow the milk and all the time trying to locate every noise and movement in the room with its great nervous eyes is one of the quaintest pictures imaginable.

(k) CURRENCY.

That the chank once served a savage people as a form of currency is little known, but so it was in the Naga country of Assam until less than 50 years ago. Major-General John Butler, who commanded an early expedition into the Naga hills, tells (loc. cit., p. 157) that he found the Nagas of many villages using chank-shells as currency with a fixed and thoroughly well-determined exchange value relative to the price of all articles of trade. Slaves and cattle in particular were always valued in chank-shells. Thus while a male slave was worth one cow and three chank-shells, a female slave—

much more valuable, the suffragettes will learn with pleasure, than a mere man—was worth as much as three cows and four or five chank-shells. Now a cow was valued at ten chank-shells, a pig at two shells, a goat was the same rate, and a fowl at one packet of salt. As a chank-shell was considered worth one rupee, a short calculation will show that a male slave was worth Rs. 13, and a female slave Rs. 34, or 34 shells. The ransoms of villages captured during raids in these good old days were largely paid in chank-shells, beads, cows, pigs and other portable wealth. Chank-shells and beads were the chief items of currency, but even in Butler's time the inevitable invasion of the rupee was already successful in the valleys most accessible to low-country traders. At the village of Hosang-hajoo the chief remarked to Major Butler, with a show of considerable pride, " since we became British subjects, we have paid revenue in coin and with it we can procure anything we require ; we therefore no longer want shells and beads."

I see no reason to believe that chank currency ever extended beyond the hill peoples of Assam and possibly some of the adjoining hill tracts. On some coins issued by the ancient Pandiyan and Chalukyan dynasties of southern India a chank-shell appears as the principal symbol (Thurston, I, 328) ; this might be held as evidence of a preceding currency consisting of the actual object so represented, whereof the memory was perpetuated in pictorial form upon one face of the coins and tokens which came to take its place as more convenient units of exchange. But there is much more reason to believe that the chank was represented on such coins for a similar reason to that which actuates the present-day States of Travancore and Cochin to adopt a similar symbol on their current coins. In these two States, the homes of southern Hindu orthodoxy, the chank-shell symbolises the religious belief of the ruling race and is their emblem as the rose stands for England and the thistle for Scotland. Both these States utilised it as a distinctive symbol on their earliest issues of local postage stamps in place of and to the exclusion of the sovereign's head—the customary pivot of design in European stamps.

Both the States of Travancore and Cochin also employ the chank in their recently designed armorial bearings. In the case of Travancore, the arms described in heraldic terms consist of :—Argent, on a fesse azure, three reversed (sinistral) chank-shells or ; Crest—a sea-horse proper. Motto—*Dharmosmat Kuladevatam.* In Cochin the shield bears more numerous devices ; in addition to a left-handed chank, a palanquin, a brass lamp and an umbrella are depicted, with elephant supporters as in the case of Travancore. In all cases where these States use the chank symbol, it is necessary to note that it should occur in the abnormal sinistral or reversed form, this being the Royal and Sacred Chank—the Chank of Vishnu.

When the Maharaja of Travancore performs *tulabharam*, a coronation ceremony wherein he weighs himself in scales against gold, special gold coins are struck called tulabhāra kāsu (*cf.* our Maundy money). On one side a figure of a chank-shell appears, on the other the legend " Sri Padmanābha " in Malayalam characters. After the ceremony these coins are distributed among the Brahmans who have assembled from all parts of the country.

(l) CHANK LIME.

A minor use to which chank-shells are put in the coastal districts where they occur, and also in those localities in Bengal where bangle factories exist, is to calcine these in kilns. The lime so produced is esteemed the best quality obtainable in India, fully equal to, if not better than, that obtained by burning pearl oyster shells. The auspicious nature of the shell adds further value to the product, and when a temple or shrine or specially fine newly-built house has to be whitewashed, chank lime is greatly sought after for this purpose in the Tamil districts. I have even received petitions praying that permission be granted for the collection of chanks for this purpose.

At the present day the fact that almost all the produce of the South Indian chank fisheries is exported to Bengal, makes it very difficult to obtain chank lime—the shells are too valuable to calcine. That it was not so in former times, in some cases at least, is to be seen if we inspect the walls of the old temples at Korkai, the seat of the Tinnevelly chank fishery 800 to 2,000 years ago. The mortar still contains many recognisable fragments of chank-shells.

(m) IN MEDICINE.

Apart from the uses to which chank rings and bracelets are put as amulets against certain ailments, the shell itself in several ways is used medicinally. Except in cases which have come under my personal notice, it is somewhat difficult to ascertain the exact nature of the diseases for which native practitioners employ this specific ; custom appears to vary with different districts and even with different " doctors " living in the same town.

Of some there is no doubt. The belief is general throughout Tamil districts and Malabar that water which has been in contact with an article formed from a chank-shell is a charm against and a remedy for blotches, pimples and other skin troubles on the face and body. A chank ring worn on a finger is an easy way of applying the remedy, as water applied to the face or body by the hand must necessarily have been in contact with chank substance and so able to transmit the virtue thereof. This remedy is believed to act still more beneficially if the ring be rubbed upon the affected parts. In South Arcot, Tanjore, Coimbatore, Salem and Trichinopoly, certain skin diseases, eruptions, warts and even hæmorrhoids are believed to yield to this treatment. In Coimbatore native doctors prescribe a paste made by mixing chank powder in water or by rubbing it up with human milk for use as a salve in the case of eruptions (sties) on the eyelids. Chank ointments (basmams) are also employed in the same district to cure inflammation of the eye, the growth of bad flesh (granulation) on the interior surface of the eyelids and also for piles and leprosy.

Chank-shell in the form of powder is also stated to be taken internally in South Arcot, Salem, Madura and Tinnevelly, either in water or mixed with ghee, as a specific for skin eruptions, asthma, coughs, and also to cool the system. In Salem and also in Ceylon it is used as a remedy for consumption. Both in Tanjore and Salem mixed with milk or water it is also employed as a salve or lotion applied to pimples and boils. In Malabar and South Canara, I am told, it is used in the case of rickets (*grahani*), chank ring powder ground in water being rubbed on the breast. At Tanur a street quack told me he used chank-shell powder internally as a remedy in cases of *varchcha* (gonorrhœa, I believe).

Among the Tuticorin Parawas a mixture of camphor and chank powder is commonly used to relieve soreness of the eyes. A small piece of camphor is partially burned and then ground down in a small quantity of human milk upon a flat stone by means of a small well-cleaned chank-shell; a small amount of powder from the shell is thus incorporated with this peculiar ointment; sometimes the white of an egg is substituted for human milk. The ointment thus made is applied round the eyelids; it is reputed to effect a sure and speedy cure.

Pounded chank-shell is also given internally by native practitioners in Trichinopoly, Salem and Coimbatore to those who suffer from an acute form of dyspepsia called *kunman*. It is administered after each meal—a treatment perfectly rational as the carbonate of lime of which the shell is composed is well adapted to counteract hyperacidity of the gastric fluids.

In Gujarat and Kathiawar chank powder is prescribed as a specific in the following diseases :—Jaundice, phthisis, coughs, shooting pain in the side, general debility and, very commonly, in affections of the eyes.

With regard to the practice of prescribing it in the case of asthma, cough and consumption, a medical friend points out that while of no value in asthma, this treatment has reason for its employment in phthisical cases—the introduction of quantities of lime into the system facilitating the deposit of lime salts around tubercular centres, encapsulating them and rendering them innocuous.

In rickets the use of lime taken internally is also indicated emphatically, the disease being characterised by an insufficient deposit of lime in the bones. He also points out that in the case of hæmorrhoids, the use of lime administered internally may assist a cure by increasing the coagulative property of the patient's blood.

It appears therefore that the employment of chank-shell powder by native practitioners is not without reason in regard to certain diseases, and while it may be objected that a non-organic form of carbonate of lime should prove equally beneficial, it has to be remembered that the carbonate of lime of shells is laid down within a delicate framework of animal membrane, and this minutely divided form may possibly render it more easy of assimilation in the body and therefore more efficacious. The religious associations surrounding the chank have also their value in inspiring the confidence of patients in the value of this medicine, faith that may help largely towards a cure. The wearing of chank rings, the rubbing of the affected parts with them and the laving of them with water which has been in contact with these rings, are forms of treatment on a different footing. They are to be considered purely as charms, without direct therapeutic value. They bear the same relation to the internal employment of powdered shell as does the quack exploitation of electricity by means of belts and bands containing discs of metal to the legitimate use of current electricity in the hands of qualified medical practitioners. If the former have any value it is by reason of faith alone.

The egg-capsule of the chank is employed by the chank and pearl divers of Tuticorin to relieve headache. They grind up a portion of the egg-capsule (sanku-pu or " chank-flower ") in gingelly-oil, together with pepper and coriander seed, and apply the paste to the forehead and temples.

Finally, according to Risley (II, p. 223) the shell-workers of Dacca are accustomed to extract the dried remnant of the visceral coil (called *pitta*) from the shells they receive and to sell this to native physicians as a medicine for spleen enlargement. He also states that the dust produced in sawing the shells is employed to prevent the pitting of small-pox and as an ingredient of a valuable white paint.

(n) FOOD.

During the run home from the chank beds, the divers are accustomed to extract the foot and anterior part of the body of the chank from the shells they have collected. The work is roughly performed by means of a pointed iron rod and all the apical mass, comprising the hepatic and reproductive glands, remains within the shell. What is extracted consists a'most entirely of tough muscular tissue carrying the adherent horny operculum at one end. These fragments are collected in the little palmyra-leaf baskets used for bailing water out of the canoe. The flesh, called *chanku-chathai*, is carried home and there prepared for family use. The preparation consists of separating the operculum, boiling the flesh for a short time and then cutting the foot and head region transversely into thin slices. These are dried in the sun ; when required for use they are fried in oil and eaten with rice and curry stuffs. On one occasion I essayed to try this much-esteemed food, but my taste was not sufficiently cultivated ; the fried shces tasted or rather smelled like frizzled shoe-leather and were altogether too tough for my teeth.

(o) INCENSE STICKS.

The horny operculum is also put to use. It is dried, reduced to powder, and then employed, after soaking in water, as an adhesive matrix to bind together the powdered adapted to this purpose.

In the Laccadive Islands all the inhabitants are required under penalty to attend the call of the chank, sounded in cases of emergency and public requirement. Among these are counted the beaching of boats and the inauguration of rat hunts.

To conclude this account of the miscellaneous uses to which the chank is put and of which the foregoing summary has by no means exhausted the list, the following instance of the ingenuity of the Indian countryman may not be amiss. For it I am indebted to Mr. C. A. Innes, I.C.S. Apropos of a flight of winged termites, he told me that once when travelling in the Madura district, he chanced upon a low caste man engaged upon some mysterious work on a large termite anthill : the man had a chank-shell in his hand. When asked what he was doing, he replied, " I am catching white ants to eat," and gave a blast upon the chank at one of the major openings into the hill. Hardly had he finished ere crowds of ants sallied forth from other openings, and these the man scooped up in handfuls and ate without any preparation.

IV.

THE USE OF CHANK BANGLES.

(1) IN NORTHERN INDIA.
(2) IN THE MADRAS PRESIDENCY.

(1) *In Northern India.*—Although evidence is strong in favour of the belief that the custom of wearing chank bracelets was in old times prevalent throughout the length and breadth of India, more especially in the Tamil country, in the Deccan, in Kathiawar, Gujarat, and Bengal, at the present day only in Bengal, the adjacent hill regions to the west, north, and east and in a few Tamil-speaking districts in the extreme south of India, does the custom continue to be widely observed and of notable social importance.

In Bengal and wherever in the adjoining Provinces of Assam, Behar and Orissa there are colonies of the Bengali race, every married woman of all castes which are thoroughly Hinduised is bound to possess a pair of chank bangles lacquered in vermilion as one of the visible tokens of her married state ; the red sankha, or shakha as it is called in Dacca, is indeed as necessary of assumption during the marriage ceremonies as is the performance of that other Hindu custom of smearing a streak of vermilion on the forehead or down the parting of the bride's hair or as the wedding ring of English women. Garcia da Orta's curious statement quoted on page 73 is to be explained in the light of this custom ; his informants doubtless meant to convey no more than that among the better classes an essential part of the marriage ceremony consisted in placing chank bracelets on the arms of the bride. The women of castes holding good social status appear, however, to have no great liking for the custom, particularly if their husbands be well-to-do, and I was informed that they frequently lay them aside temporarily in favour either of more handsomely carved ones or replace them when means permit by gold and jewelled ones. Chank bangles are occasionally ornamented with gold and set with jewels ; the price of these may reach several hundreds of rupees. The great majority of married women, however, wear them permanently, never removing them so long as their husbands are alive. Occasionally some of the modern sankha (marriage) bangles are made in two sections secured together after the bangle is placed on the wrist by means of tiny bamboo pins as it is otherwise impossible to pass one of the right size over the hand without great difficulty and the infliction of acute pain.

In spite of the rapid spread of a desire for bracelets of more showy appearance, there are very large numbers of prosperous Hindu households, especially in the country districts, where the womenfolk remain attached to the old and less ostentatious custom of wearing chank bangles solely as ornaments. Among these conservative folk a large demand exists for the handsome products of the sankhari workshops. The ornamental bangles made to meet these requirements are of two kinds called respectively *bála* and *chári*. The former are broad bangles worn one on each wrist. The *chári* on the con-

TEXT-FIGURE 1.

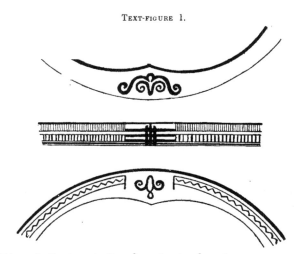

Edge and side views of a Bengal marriage bangle to show ornamentation
in yellow upon a red-lacquered ground. From Pabna, Bengal. × 2.

trary is always quite narrow, generally $\frac{1}{6}$ to $\frac{1}{5}$ inch in width, and usually of conventional scroll design, worn in a set of three on each wrist.

The use of these ornamental bangles (*bála* and *chári*) and also of the red marriage bangle is limited almost entirely to the thoroughly Hinduised sections of the Bengali people, together with the Hindu communities settled in Assam, Behar and Orissa. Baishnab women, however, do not wear these bangles according to the Collector of Birbhum. Information received from a Muhammadan source indicates that women of the lower classes of this community in Dacca, Darjeeling and Assam occasionally wear chank bangles as wrist ornaments.

As elsewhere in India, it is the invariable custom in Bengal in orthodox Hindu households for widows to discard all their jewellery on the death of their husbands. In the case of chank and glass bangles, it is usual for the widow to break and throw them away on the first occasion when she bathes after her husband's death. They never resume the use of similar bangles except in the very rare cases where re-marriage is permitted to widows. Tavernier says[1] "when a man dies, all his relatives and friends should come to the interment and when they place the body in the ground they take off all the bracelets which are on their arms and legs and bury them with the defunct." This burial of the widow's bangles with the dead may still be continued by some castes, but as earth-burial is now rapidly being displaced by cremation as orthodox Hinduism secures a firmer hold on the people, this custom must tend to die out. Generally in Bengal the Hindu women wear sankhas as visible tokens of the possession of living husbands. The Hindu Shastras are said to enjoin their use as it is believed that this contributes to the prosperity and longevity of their husbands.

Surti, Tuticorin and Rameswaram chanks are necessary in the manufacture of both *bāla* and *chāri* bangles as these require to be made from the finest quality of shells —those possessing a pure white porcellaneous appearance and a dense well-conditioned substance susceptible of high polish.

Among Bengal castes of inferior social status, particularly those whose physical characteristics bespeak Dravidian descent and whose customs are not yet thoroughly Hinduised the use of chank bangles made up into massive gauntlets composed of numerous separate bangles is very prevalent. Prominent among these are the widely spread Kochh tribe in their two principal sub-divisions of Rajbansi and Paliya. It is largely to supply the women of this tribe with their characteristic ornaments that the chank bangle workshops in Dinajpur and Rangpur exist, as in these districts the tribe has its chief settlement with an approximate total of one million individuals. Kuch Behar and Jalpaiguri account for another half million, while considerable numbers are found also in Purnia, Maldah, the Darjeeling Terai, Bogra, Murshidabad, Nadiya and Dacca. The Rajbansi and Paliya gauntlets are composed usually of ten separate bangles. As the wearers belong largely to the labouring and agricultural classes the bangles forming these gauntlets are broad and thick, frequently without any ornament whatever; where decoration is attempted, it consists of simple line patterns made of shallow groovings which impair very little the strength of the bangle and yet are very effective and elegant (Text-figure 2). Neither are they usually polished, hence dead shells from Jaffua are largely employed in this manufacture, although inferior shells of the better qualities from the Indian side are also extensively made use of.

The Muchi is another important Bengal caste where the wearing of numerous chank bangles is a distinctive custom among the women. This is a leather-dressing and cobbler caste, socially a shade higher than the allied Chamārs from whom the Muchis appear to be an offshoot. One of the obvious distinctions between the women of these castes lies in the character of the bracelets worn. Thus while the female Chamār prides herself on huge bracelets of bell metal adorning her arms, the Muchi woman always substitutes chank bangles. The Muchis, like the Paraiyar of the South, are largely

[1] *Loc. cit.*, Vol. II, p. 285.

the caste drummers of the province and as they are fond of the violin and the pipe are usually employed as musicians at Hindu weddings.

In Western Bengal and in Behar the Santáls take the place occupied by the Rajbansis and Paliyas in North-Eastern and Eastern Bengal as the chief chank bangle wearing tribe.   Many of their women follow the same habit of disposing of a number of chank bangles, three to five usually, as a massive cuff-like gauntlet or compound bracelet. These people being generally poor, the quality employed for these compound bracelets is inferior and red and yellow lac are freely used upon them to enhance their appearance and to disguise imperfections.   Many indeed are too poor to afford these ornaments and others belong to families which do not observe the custom ; in Birbhum which

TEXT-FIGURE 2.

Gauntlet pattern of compound bracelet worn by Paliya women, Bengal.

may be taken as a characteristic Santál district, it is estimated that about half the female Santál population follow this custom.   Sometimes Santál girls wear them from an early age but generally they are assumed at marriage.   It has no religious significance and marriage may be performed without the putting on of these bangles which are worn more for ornament and because of custom than for any more serious reason.   Like the Hindus, Santál women break and throw away their bangles on the occasion of widowhood, re-assuming others, if they wish, if they remarry.   Dead shells are often employed by the cutters for Santál bangles.

Risley states (II, p. 225) that the Santáls in point of physical characteristics may be regarded as typical examples of the pure Dravidian stock and in view of the similar origin attributed to the Kochh tribe which includes both the Rajbansi and the Paliya, this becomes a matter of great significance as well as of much difficulty, for whereas the Kochh people are professed Hindus, the Santāls hold the animistic beliefs characteristic of Non-Hinduised Dravidians. However Oldham, as quoted by Risley (I, p. 492), states that " the adhesion of the Kochh tribe to Hinduism is comparatively recent as shown by their own customs as regards burial, food and marriage."

The section of the Kurmi caste found in Chota Nagpore and Orissa also wear chank bangles. In view of what has been said above in regard to the Dravidian origin of the Kochhs and Santāls, it is of importance to find that Risley (I, p. 530) considers this territorial section of the caste as undoubtedly Dravidian, as shown by their physical characteristics, religious beliefs and social customs. In appearance, he says that in Munbhum and the north of Orissa, it is difficult to distinguish a Kurmi from a Bhumij or a Santāl. In their religion the animistic beliefs characteristic of the Dravidian races are overlaid by the thinnest veneer of conventional Hinduism, and the vague shapes of ghosts and demons who haunt the jungles and the rocks are the real powers to whom the Kurmi looks for the ordering of his moral and physical welfare.

Alike with the Santāls the internal structure of that branch of the Kurmi caste living in Chota Nagpur and Orissa is founded upon a distinct and well-defined totemism in which a large proportion of the totems are still capable of being identified. Risley (II, appendix, p. 88) enumerates 60 totemistic sections or septs in this caste, among which is one termed Sankhawār whose members are prohibited from wearing chank shell ornaments. Among the Santāls, the place of this sept is taken by one called Sankh, wherein all individuals are forbidden, under pain of caste punishment, the use of the chank shell in any form ; they may neither cut, burn, nor use the shell, nor may the women of this sept use it in personal adornment (I, p. xliii).

The prevalence of the use of chank bangles among these Dravidian races, the present animistic beliefs of the Santāls and Chota Nagpur Kurmis, and the comparatively recent renunciation of the same cult by the great Kochh tribe, taken in conjunction with other facts and especially with the widely spread archæological finds detailed elsewhere in these pages, point to the use of chank bangles as having had a purely Dravidian origin and as having been a custom prevalent and solidly established among at least certain sections of the race throughout India anterior to the advent of the Aryan invaders and the rise of the Brahmanic faith. The cult of the chank would therefore appear to be one adopted (and modified) by the Brahmans from the religious belief which they found indigenous to India.

Finally, in the hill tracts of Chittagong, we find the women of the Maghs, a race of Indo-Mongolian extraction and Buddhists by religion, using very broad unornamented sections of chank shells as bracelets in similar manner as we shall next see is the habit in Thibet and Bhutan, inhabited by other Mongolian races. To supply the needs of the Maghs, bangle cutters are established in Chittagong ; these work-people are chiefly Muhammadans and the work they do is of the roughest and crudest description in conformity with the undeveloped artistic taste of their customers who appear to wear these bracelets rather as amulets than as ornaments. Broad arm ornaments of similar simple form are used by the Papuans and by the wild inhabitants of several groups of the Melanesian islands ; sometimes round the wrist, sometimes on the upper arm above the elbow. I do not know, however, whether the shell employed in these instances be Turbinella or not. Among these island tribes it is the men who wear these ornaments.

Outside of Bengal and Assam the only considerable demand for chank bracelets comes from Thibet and Bhutan. The trade is one of long standing, for Tavernier, in 1666, found Bhutanese merchants taking home from Pabna and Dacca bracelets sawn from " sea-shells as large as an egg." He also states that 2,000 men were occupied in these two places in making tortoise-shell and sea-shell bracelets and " all that is produced by them is exported [1] to the kingdoms of Bhutan, Assam, Siam and other countries to the north and east of the territories of the great Moghul " (*loc. cit.*, p. 267).

Chank bangles appear to be worn very generally throughout Thibet, from Ladakh in the west to the Kham country in the east. Neve records [2] seeing the poorer women in Kashmiri Thibet wearing broad shell-bangles in shape like a cuff on both wrists, while on the march of the British expedition to Lhassa in 1904 they were noted as in frequent use by Thibetan women. This ornament is assumed early in life while the hand is still small and pliable ; after a few years it becomes impossible to remove it without breakage which these women will suffer only in the last resort, as it cannot be replaced except by one of large diameter which will fit more loosely on the arm than they like. A medical officer with the Thibet mission has informed me that in one instance a Thibetan woman was brought to him for the treatment of a festering wound on the wrist. On examination the cause of the trouble was found to be the presence of a chank bangle so small that the wrist had been wounded and circulation impeded ; gangrene was imminent and although the woman was loth to part with her bangle it had to be filed off to save the hand.

The export of round and square discs of chank shell to the Buddhist countries of the north appears to be much less than in Tavernier's time, as it is now comparatively insignificant. Among the Nāgas, the discs are employed to ornament the men's hair-bedecked helmets, and Prince Henri d'Orléans [3] found the women of the wild Lissus,

---

[1] Evidently a *lapsus calami* as the custom of wearing chank bangles was even more prevalent in Tavernier's day among Bengali women than it is to-day, *vide* Orta, *loc. cit.*

[2] " Beyond the Pir Panjal," London, 1912.

[3] " From Tonkin to India," English Translation, London, 1898.

a section of the Lolo tribe, mountaineers living in the upper valley of the Mekong in Yunnan, employing chank shell discs to ornament their Chinese caps. It may be that these Lissus and cognate tribes represent those chank jewel wearers whom Tavernier refers to as belonging to the kingdom of Siam. In this latter country at the present day I know of no utilisation of chanks in personal adornment.

The chank is one of the eight lucky signs recognised by Buddhists of the Northern cult and as such is constantly reproduced in Buddhist ornamentation in Thibet and Bhutan.[1] It may therefore be inferred that the use of it in personal adornment has a like reason ; whether in the form of a bangle, a cap or a hair ornament, a necklace or a breast disc, it is employed as a talisman to ensure good fortune, and possibly even as an amulet against the evil eye, as is the chank shell placed on the forehead of draft bulls in Southern India.

(2) *The Tribes and Castes which wear Chank Bangles in the South of India.*

In the Madras Presidency and the associated native States, the castes whose women systematically wear chank bangles are few, and if we except the wandering tribes of the Lambadis (or Brinjaris), Koravars and Kurivikkarans, the custom appears confined to a sub-division of each caste or tribe. Whether it had a totemistic origin and significance as it still has among non-Hinduised tribes in Bengal, Behar and Chota Nagpur is not at present clear. If it had, the original tribal sept, usually exogamous, has become changed to a caste sub-division, invariably endogamous. And whereas among the septs of those animistic tribes in Northern India which are named after the chank this shell is taboo with them, it forms the characteristic ornament of the women of the caste sub-divisions named after it, in Southern India.

Only in the Kongu country, which coincides roughly with the present inland districts of Coimbatore and Salem, does the custom continue to flourish at all strongly. Coimbatore is the great centre of the custom, for there the numbers of chank-bangle wearers greatly exceed those found in any other district. The Collector reports eleven castes and sub-divisions as following this custom, viz. :—

(1) Pala Vellalas.
(2) Puluvans (so-called Puluva Vellalas).
(3) Konga and Golla Idaiyans.
(4) Konga Shanans.
(5) Konga Vannans.
(6) Thotti Chukkiliyans.
(7) Sangu, Konga, Sangudu or Sanguvalai Paraiyans.
(8) Thottiya Naiks.
(9) Okkilians (not universal).
(10) Kurumbars.
(11) Lambadis in parts of Kollegal and Gobichettipalaiyam divisions.

---

[1] J. Claude White, "Sikhim and Bhutan," p. 46. London, 1909.

The custom is associated particularly with those caste sub-divisions whose territorial cognomen indicates a long settled residence in the Kongu country (Coimbatore and Salem). The Konga sub-divisions of the Idaiyar, Paraiyar, Vaunar and Shanar castes have this custom in common and as several other castes in Coimbatore also adhere to it, we may infer that at one time the custom was general in the Kongu country among at least the generality of the lower castes.

Another caste sub-division where the women wear chank bangles is that of the Sangukatti Idaiyans. Among them the marriage ceremony requires, as in Bengal, the placing of a chank bangle on one of the bride's wrists.

Among Coimbatore castes the chank bangle is worn always upon the left wrist, usually singly but occasionally a pair in certain cases, e.g., among the Konga Paraiyans.

The wearing of these bangles is considered as a symbol of the permanence of the marriage tie, a belief probably derived from the custom of breaking the bangle after the death of the husband. The Collector of Coimbatore states that a widow discards her bangle one month after her husband's death. He adds that if a woman accidentally breaks her bangle, she thinks it unlucky and regards it as an omen that her husband will chance on some evil ; when the husband is sick the wife prays that it may be her good fortune to wear the bangle during her whole life. A woman considers it improper to appear before her husband or in public without the bangle. The wearing of this ornament appears to be followed in Coimbatore largely because it is an ancient custom, with no further significance beyond what is implied above. It is not now connected with belief in the evil-eye though it is said to have had this significance in former times. The general belief in the efficacy of the chank as a specific against skin diseases may however be counted as one of the obscure reasons for its continued usage.

The practice is more general among low castes. The Collectors of Madura and Trichinopoly both inform me that among Paraiyans, Chukkiliyans (leather workers, etc.), Oddans, Koravas and the Naick sections known as Kavaraja and Thottiya Kambalathans, together with the wandering tribe of Lambadis, the custom of wearing chank bangles is found to prevail here and there in both districts. There appears to be no general observance of the custom—in some villages and taluks none among the women of the castes named wears chank bangles ; elsewhere, as in the Nāmakkal Taluk (Trichinopoly District), a definite section of the Paraiyan caste called Sengudimi Paraiyans adopts this ornament as a distinguishing sept distinction, while in other parts of the country, the women of these various low castes wear it chiefly if not entirely for its ornamental value. The custom appears to be dying out, as witness the vagueness of the people who still adhere to its observance as to the reason for so doing, its partial and sporadic geographical distribution in the districts where it lingers, and the comparatively small numbers who adhere to it. As a typical instance of the irregular distribution of the custom, the report of the Collector of Trichinopoly states that in the Musiri and Kārūr Taluks, no people wear chank-bangles; in Nāmakkal Taluk only the Paraiyan sept call Sengudimi Paraiyans wear them; in Udaiyārpālaiyam Taluk the habit is confined to Koravans and Lambadis; in Trichinopoly and Perambalūr Taluks to Chukkiliyans and Paraiyans, while in Kulittalai Taluk besides the two castes just named, the Oddans, Koravans and Thottia Naickers are given as castes following the custom—it is to be noted that all are amongst the lowest of castes, constituting what may be termed the inferior labouring population. The bangles employed are manufactured in Kilakarai and pass to Trichinopoly by the intermediary of traders in Madura. The price in Trichinopoly town is from Rs. 3 to 4 per 100 bangles, all very roughly made, with little or no ornamentation.

In Tanjore and Salem the Paraiyans and Chukkiliyans use chank-bangles in certain villages and these Salem Pariayans are said to belong to the Konga division as in Coimbatore. In both districts the Lambadis and Koravas have the same custom, and the Collector of Tanjore adds that the women of the Uppu Koravar, Panni Koravar, and Vari Koravar sub-divisions, together with the Oddar and Domba castes, all agree in following the habit. He states that the Koravas put on the bangles during the marriage ceremonies.

In Salem the Malayali women of the Chitteri Hills, Uttangarai Taluk, also use chank-bangles.

Hawkers called Dasam Chettis, who bring their wares from Rāmēswaram and Kilakarai, attend village fairs and temple festivals with these bangles in Salem and Coimbatore, charging from 4 to 12 annas a pair. In Coimbatore well-to-do Puluva Vellalans not infrequently wear bangles of superior quality costing even Rs. 3 to 5 per pair; these probably are brought from Calcutta, as no expensive patterns are made at Kilakarai. In Tanjore about the same prices prevail, but here the most expensive, said to come from Calcutta, are said to cost one rupee per pair; common qualities as usual come from the Rāmnād coast.

In the Nilgiri hills, especially in Ootacamund, Konga Paraiyan women who have come from Coimbatore are often to be met with wearing chank-bangles. Their sub-division is considered one of superior standing and the people belonging to it will not act as sweepers. The bangle is always worn on the left wrist; frequently two are worn, always plain and massive, and about ⅜ inch in width, exactly similar indeed to those worn by Chanku (Puluva ?) Vellalas.

The women of the Kota hill-tribe in the Nilgiri hills have an allied custom, but instead of a massive bangle cut from the entire shell, they wear around the left wrist a bracelet of roughly-made chank beads strung on a thread. When chank beads are not obtainable they wear a string of white glass beads as near the colour and shape of the usual chank beads as they can obtain. In answer to my questions, the elders of a Kota village situated close to Kotagiri informed me that all married women must possess and wear one of these bracelets on the left wrist together with two massive copper bangles on the right wrist during their husband's life. They are assumed, however, before marriage, and the putting on of the chank bracelet is not a part of the marriage ceremonies. On the death of the husband the widow discards this with her other ornaments, but is permitted to resume them after a decent interval of mourning—three months according to my informant. It is to be noted that the wife's ornaments are not broken or destroyed at the husband's death as is the custom among the chank-wearing Hinduised plains-people. At the woman's own death they are put to burn with her body on the funeral pyre. The Kotas can adduce no special reason for the wearing of these bracelets, except that their god Bhagawani long ago ordered that their women should do so.

From the foregoing it is seen that chank-bangle wearing is confined to the Tamil districts in the south of the Madras Presidency. It is unknown in the central and northern sections—the Madras Deccan and the coastal Telugu districts—except in respect of the wandering Lambadis ; the Collector of Kurnul informs me that chank-bangles for sale to the local representatives of this tribe are occasionally brought from the Raichur side (Hyderabad State), a significant fact as the Raichur Doab is one of the localities where Mr. Bruce-Foote found numerous fragments of chank factory waste, indicative of the former existence of a bangle-making industry in the vicinity of his discoveries. Where these modern Raichur bangles come from I do not know, but I should expect them to be of Bengal manufacture.

(1) THE ANTIQUITY OF THE CHANK-BANGLE INDUSTRY.

(a) IN THE TINNEVELLY DISTRICT.

Reference to ancient Tamil classics furnishes evidence scanty but conclusive of the existence of an important chank-cutting industry in the ancient Pandyan kingdom in the early centuries of the Christian era. Similar evidence is also extant of a widespread use of carved and ornamented chank bangles in former days by the women of the Pandyan country which may be considered as roughly co-extensive with the modern districts of Tinnevelly, Madura, and Râmnâd, forming the eastern section of the extreme south of the Madras Presidency.

Among the more important references which prove the ancient importance of this industry on the Indian shore of the Gulf of Mannar, is one contained in the " Maduraikkanchi," a Tamil poem which incidentally describes the ancient city of Korkai, once the sub-capital of the Pandyan kingdom and the great emporium familiar to Greek and Egyptian sailors and traders and described by the geographers of the 1st and 2nd centuries A.D. under the name of Kolkhoi. From the purity of the Tamil employed in this poem and the similarity of the names of the towns, ports and goods mentioned incidentally with those employed by Ptolemy and the author of the " Periplus of the Erythræan Sea," we may date it as approximately contemporaneous with the writings of these authors and certainly not later than the 2nd or 3rd century A.D.

In one passage (LL. 140-144) the Parawas are described as men who dived for pearl oysters and for chank shells and knew charms to keep sharks away from that part of the sea where diving was being carried on. Another passage depicts the city of Korkai, then a seaport at the mouth of the Tambraparni, as the chief town in the country of the Parawas and the seat of the pearl fishery, with a population consisting chiefly of pearl-divers and chank-cutters. The great epic, the *Silappathikkarram*, or " Lay of the Anklet," written about the same period by a Jain poet, gives further information about Korkai, from which we gather that on account of the great value of the revenue derived from the pearl fishery, this city was a sub-capital of the Pandyan realm and the usual residence of the heir-apparent, boasting great magnificence and adorned with temples and palaces befitting its wealth and importance.

Another valuable reference to the chank trade is contained in two Tamil stanzas which chronicle a passage at arms between a Brahman and Nakkirar, the celebrated poet-president of the Madura Sangam in the reign of the Pandyan king Nedunj Cheliyan II, who flourished probably about the beginning of the 2nd century A.D.

The Brahman, named Dharmi, presented to the Sangam a poem purporting to be composed with the aid of Siva. Nakkirar, the President, in spite of its alleged divine origin criticised the poem mercilessly, and rejected it as unworthy of literary recognition. The Brahman took revenge by presenting another poem also purporting to be inspired by Siva ; in it he held the President up to ridicule on account of his caste trade in pungent lines which may be translated literally as follows :—

" Is Kiran fit to criticise my poem ? Spreading his knees wide, his joints loosened (by the labour) does he not saw chanks into sections, his ghee-smeared saw murmuring the while kir—kir ? "

Besides the insult intended to be given, the verse contains a play on the President's name and the sound given out during the sawing of chank shells.

The reply of Nakkirar was " Chank-cutting is indeed the calling of my caste ; of that I am not ashamed. But of what caste is Sankara ? (one of the many names of Siva). We earn our livelihood by cutting chanks ; we do not live by begging as he did "—an allusion to the fable popularised by the Brahmans wherein Siva is represented as a mendicant seeking alms with a skull in his hand as begging bowl.

Dharmi's description of a chank-cutter's trade is wonderfully vivid in the original Tamil ; in a dozen words he paints a realistic word-picture of a cutter's workshop—the men seated on the ground with the knees widely spread and depressed outwards almost to the ground to give free play to the great crescentic two-handled saw monotonously droning a single note as it cuts its way laboriously through the hard substance of the shell.

Tradition has it that Nakkirar, the chank-cutter President of the Sangam, was a Parawa by caste. It would be most appropriate if this be correct as we have already seen that at the beginning of the Christian era chank-fishing and chank-cutting were among the important trades carried on in Korkai, the chief settlement of the Parawas in early days.

No Parawas to-day are engaged in chank-cutting although they still largely monopolise the shore industries of Tinnevelly where they continue as from time immemorial to provide the contingent of divers required for the exploitation of both the pearl- and the chank-fisheries of the Gulf of Mannar.

It is noteworthy that though their women do not now wear chank-bangles their children from four months to about two years old are often given roughly-made chank-bracelets to wear in the belief that such will protect them against the baleful influence of the evil eye, from vomiting and from a wasting disease called *chedi* which appears to be rickets and reputed to be caused by the touch or near approach of a woman during her menses ! This custom has now been abandoned or is perfunctorily performed by some of the better class Parawas, but the great majority, including naturally the whole of the poorer and more ignorant sections of the community, continue to adhere strongly to the custom. The bangles are roughly fashioned and with the crudest of ornamentation ; they are made by Muhammadans at Kilakarai, their chief settlement on the coast of the Gulf of Mannar.

Apart from this evidence we have nothing of importance till we come to the sixth century when the travelled monk, Cosmas Indico-Pleustes, after mentioning the island of Ceylon, proceeds to say " and then again *on the continent* and further back is Marallo which exports conch shells (κοχλίους)." Sir J. Emerson Tennent in his " Sketches of the Natural History of Ceylon " (London, 1861) misses the significance of the expression " on the continent " and identifies Marallo with Mantotte near Mannar on the north-west coast of Ceylon, where chanks are collected in the neighbourhood in large quantities even at the present day. Yule [1] with closer adherence to the old text would place this ancient chank-fishery on the Indian coast (*i.e.*, on the continent opposite Ceylon), and he suggests that Marallo is a corrupted form or misrendering of Marawar, the name of the chief caste living in the coastal district of Rāmnād, now the location of one of the most productive and accessible present day chank-fisheries. The name of the local people not infrequently was applied by old travellers to the chief town in their territory and so, very reasonably, we may identify Marallo generally with the Maravan coast and particularly with either the town of Rameswaram or of Pamban situated at the western extremity of Adam's Bridge and directly opposite to Mantotte and Mannar at the western extremity.

---

[1] *Cathay and the Way Thither*, Vol. I, p. clxxviii, London, 1866.

Ma'bar or Maabar, the Arab name for the western coast of the Pandyan country, has probably a parallel derivation, Maabar being indeed a very fair rendering by guttural Arab lips of the Tamil term Maravar.

The next writer to mention the chank is the Arab Abouzeyd, who in 851 A.D. stated that " they find on the shores of Ceylon the pearl and the shank, which serves for a trumpet and which is much sought after." [1]

A long gap occurs in references by travellers to chank-fisheries till the days of the Portuguese and Dutch when they became fairly frequent. A few years before the establishment of the former power in the Gulf of Mannar, the traveller Barbosa visited the old town of Kayal, and from him we learn that it was then still an important seaport where many ships from Malabar, Coromandel and Bengal resorted every year to trade with the rich Hindu and Muhammadan merchants living there, a definite statement which shows that there was even then no difficulty in forwarding supplies of chanks direct by sea to the Dacca workshops.

Barbosa also tells us that at the time of his visit the fishery off this coast belonged to the king of Koulam (Quilon in the southern part of Travancore) who generally resided at Kayal and who farmed the pearl-fishery to a wealthy Muhammadan.[2] The chank-fishery so far as we know has always been an adjunct to the more romantic pearl-fishery and must almost certainly have been treated in a similar manner, both fisheries being considered everywhere in India from immemorial times as prerogatives of the sovereign. About 1524, the Portuguese seized the Tinnevelly pearl-fishery, stationing a factor and guard boats on this coast—the Pescaria or fishery coast as it soon became termed. In 1563, Garcia da Orta speaks of the trade with Bengal having declined owing to the unrest caused by Muhammadan invaders in that country, but in 1644, Boccaro in a detailed report upon the Portuguese ports and settlements in India records that a large quantity of chanks fished off Tuticorin were exported to Bengal " where the blacks make of them bracelets for the arm." He adds rather quaintly the name of another Tuticorin production—" the biggest and best fowls in all these eastern parts." [3] Exactly how the Portuguese conducted this trade and what profits it yielded them are not known to me ; the Dutch, so far as they were able, destroyed the Portuguese archives in Tuticorin as well as in Ceylon, and we must await further research among the records at Lisbon before we can gain any further information.

The Dutch, keen to distinguish the substance from the shadow, paid great attention to the development of the chank-fishery as distinguished from the pearl-fishery whereof one of their most able local Governors, Baron Van Imhoff, once queried (1740) whether the latter " is not more glitter than gold as so many things are which belong to the Company, which shine uncommonly but have no real substance." [4]

[1] *Fide* Yule's " Hobson-Jobson."   [2] *Fide* Yule's " Hobson-Jobson," article " Chank."
[3] *Fide* Yule, "The Book of Ser Marco Polo," Vol. II, p. 307, London, 1871.
[4] Hornell, " Report to the Government of Madras on the Indian Pearl Fisheries in the Gulf of Mannar," Madras, 1905.

In 1700, Father Martin, a Jesuit Missionary, wrote (*Lettres Edifiantes*, II, p. 278, edition of 1843), " It is scarcely credible how jealous the Dutch are of this commerce. It is death to a native to sell them to any one but the servants of the Company. The shells are bought by the Dutch for a trifle, and then despatched to Bengal, where they are sold at great profit. These shells, which are round and hollow, are sawn and fashioned into bracelets equalling the most brilliant ivory in lustre. Those fished on this coast (Tinnevelly) are extraordinarily abundant ; they have their spiral from right to left, but if one be found twisted in the other direction, it is a treasure valued by Hindus at an extravagant price, for they believe that it was in a chank of this description that one of their gods hid himself in order to escape the fury of enemies pursuing him in the sea."

With the transfer to the British of all Dutch ports on the Coromandel coast and in Ceylon together with the acquirement of the Tanjore and Carnatic territories about the same time, the control of all the chank-fisheries in these localities passed to the British.

The evidence furnished by the Tamil classics of the existence of an extensive chank-bangle industry in the extreme south of India during the height of ancient Tamil civilisation 1,200 to 2,000 years ago has received unexpectedly conclusive corroboration within the present year (1912) through discoveries which I have made on the sites of the once famous Tamil cities of Korkai and Kayal (now Palayakayal). These cities are now represented by mounds of rubbish adjacent to villages still bearing the appellation of their celebrated predecessors. The greatest find was at Korkai, which as already noted flourished from a date well antecedent to the Christian era down to some indeterminate date prior to 1000 A.D. when the accretion of silt at the mouth of the Tambraparni drove the inhabitants to build another city (Kayal) at the new mouth of the river. Here, on the landward outskirts of the village, I unearthed a fine series of chank workshop waste—seventeen fragments in all. The whole number were found lying on the surface of the ground in a place where old Pandyan coins have from time to time been discovered according to information gathered in the village. The fragments unearthed all bear distinct evidence of having been sawn by the same form of instrument, a thin-bladed iron saw, and in the same manner as that employed in Bengal at the present day. Eight fragments represent the obliquely cut " shoulder-piece," six consist of the columella and part of the oral extremity of the shell and the remaining three are fragments of the lips—all show a sawn surface, the positive sign of treatment by skilled artisans.

At Kayal or Palayakayal (*i.e.*, old Kayal) as it is now termed, the daughter city of Korkai, which flourished in the days of Marco Polo and appears to have grown rich as Korkai gradually passed away as a sea-port owing to physical changes in the delta of the Tambraparni, I found an excellently preserved sawn shoulder-piece, with marks of the apex having been hammered in after the present-day habit in Dacca workshops.

This was found on the surface in an open space within the present village. Time did not allow me to prosecute a detailed search, but in my own mind the single fragment found is conclusive evidence of the industry having once been located here. No shell cutting of any description is carried on anywhere in this neighbourhood.

Again, at Tuticorin, I have found a sawn and hammered shoulder-piece of typical form, hence as the three discoveries were all made at places which in turn have been the head-quarters of the chank-fishery, I am fully convinced that at all three, chank-bangle workshops formerly existed, to treat on the spot this product of the neighbouring sea. Why the seat of the bangle-cutting trade became transferred or limited to Bengal is obscure, and may never be satisfactorily elucidated ; I am, however, inclined to suggest the hypothesis that the decay of the industry in Tinnevelly may have been consequent upon the Muhammadan invasion. The date of the passing away of the chank-cutting industry I am inclined to put tentatively at about the fourteenth century, a time which marks the close of unchallenged Hindu supremacy in the south, the spoliation of the vast riches of the Pandyan cities by the Moslem and the heyday of Arab sea-power on this part of the Indian coast. With the depression and decay entailed by the loot and ruin of their enormously wealthy temples and long prosperous cities by the invaders under Malik Kafur and his lieutenants it is far from improbable that the particular trade here referred to became disorganised within the Pandyan realm and forced into a different channel, the whole of the shells being exported to Bengal to be cut there instead of being treated locally at the seat of the fishery.

It is also noteworthy that the huge funeral urns found in tumuli of the Tambraparni valley (at Adichanallur) have yielded a few·fragments of working sections cut from chank shells, associated in the urns with beautifully formed bronze utensils, iron weapons and implements and gold fillets. So old are these tumuli that they are classed as prehistoric though it is obvious that the people of those days were skilful artizans in gold, bronze, and iron and must have been contemporaries of historic periods in the story of Egypt and Mesopotamia.

In this connection, as bearing upon the antiquity of the trade connections between India and Persia, some interesting exhibits are to be seen in the Louvre, Paris ; the most noteworthy is a chank-shell cup, probably used in libations, found by the Mission Dieulafoy in the ruins of Susa. It measures $6\frac{1}{2}$ inches in length and has been longitudinally bisected a little to one side of the median line, the larger portion being retained to form the cup. The columella has been cut out and the result gives a very serviceable form of spouted cup. It is classed as of Achæmenid age—say circa 500 B.C. From the same ruins the Mission de Morgan brought back a fragment of a large chank bangle nearly one inch wide (No. A7532) roughly ground to a triangular ridge pattern in section and with a rough < mark engraved at one side.

Again the Fouilles de Tello yielded to the Mission de Sarzec a number of perforated chank-shell fragments sometimes very finely engraved ; one·in particular (No. 221) bears a spirited representation of a lion leaping upon the back of a bull.   These objects were apparently ornaments worn suspended from a cord.

As there are no chank-fisheries in the Persian Gulf, these objects must have been imported from India either from Kathiawar or from the Gulf of Mannar.   The libation cup would seem to show a much more widely spread use of the chank shell in ancient times than we have hitherto suspected and research directed to this special point in collections coming from Babylonian, Assyrian and ancient Persian sites should throw further light on this subject.

(b) THE FORMER EXISTENCE OF BANGLE FACTORIES IN THE DECCAN AND IN GUJARAT AND KATHIAWAR.

I have been unable to obtain any evidence from ancient Indian writings of the existence elsewhere than in the extreme south of the country of any ancient custom of wearing bangles cut from chank shells.   Probably such references do exist, and if this be so, I trust the present notes may elicit their quotation by scholars who are familiar with the ancient Sanscrit and Gujarati classics, the most probable sources of information.

Fortunately, in this apparent absence of written records, archæology has important evidence to offer, and although it is difficult to date the greater portion of this testimony with any exactitude, it offers irrefutable proof that the industry of chank-cutting and the custom of wearing chank bangles had once much less restricted geographical range than at the present day.

Apart from the finds in the Pandyan country, mentioned above, a considerable amount of archæological evidence exists proving that extensive chank-bangle factories were located in ancient times in the Deccan and in Gujarat and Kathiawar.   The principal finds proving this were made by the late Mr. Bruce Foote ; they are now deposited in the Madras Museum and form an exhibit of wonderful and unique interest.

The remains obtained in these localities consist of two distinct series ; one comprising fragments of finished chank-bangles which appear to have been once actually worn, and discarded owing to accidental breakage or else purposely broken as a sign of mourning, the second, either of sawn rings of chank shells—working sections—ready for the bangle carver and polisher or of waste material from chank-bangle cutters' workshops.

In the Southern Deccan and neighbourhood chank-bangle fragments have been found in the following districts and States, namely :—

Mysore.
Bellary.
Anantapur.
Kurnul.
Raichur Doab.
Guntur.
Kistna.

Careful examination of the museum collection leads me to believe that shell-bangle factories existed at four centres in the area indicated, the principal being :—

(1) Hampasagra on the Tungabhadra, Bellary District.

(2) Bastipad on the Hindri River, Kurnul District.

(3) Maski in the Raichur Doab (Hyderabad State) ;

and (4) at the great Buddhist ruins of Amaravati in Guntur District.

Of all these, the last named discovered by Mr. Rea, lately Archæologist to Government, is probably by far the most important, as we are better able to date the remains here than anywhere else owing to their association with buildings of known origin.

The chank fragments discovered at Amaravati are very numerous and important, giving evidence that the chank-bangle industry was carried on in this locality even earlier than the construction of the Buddhist buildings. They consist in the main of large numbers of fragments of working sections of the shell together with a very few pieces of finished and ornamented bangles. Besides these are numerous waste pieces—shoulder and oral rejects, showing that the methods of cutting up were identical with those of the present day. The whole of these fragments were found beneath the foundations of buildings which the most competent authorities date *circa* 200 B.C., hence these bangle fragments are antecedent thereto and must be over 2,100 years old. It may be that these remains constituted part of the town's rubbish heap before the erection of the Buddhist buildings which have survived to the present time and that this rubbish heap was employed in making or raising the ground prior to the putting in of the foundations, or it may be that an old village site, including the waste of a village bangle factory, was selected as a site by the Buddhist architects of Amaravati.

Besides bangle fragments, a few rudely carved chank finger-rings figure among the remains, together with small discs of $\frac{3}{4}$ inch diameter sometimes perforated in the centre ; the latter were used in the fashioning of necklaces—at Peddamudiyam in Cuddapah, Mr. Rea has found complete necklaces formed of spherical chank beads alternating with chank-discs of the pattern here referred to. The perfectly circular outline of these small discs is remarkable.

The two fragments of chank bangles found in the Kistna District are probably of approximately the same age as those from Amaravati as they also were associated with Buddhist remains. The finds in all the other districts cannot be placed in time in spite of their frequent association with neolithic weapons and implements. Mr. Bruce Foote has indeed suggested that finely worked serrate and biserrate chert and agate flakes found in one place in association with bangle fragments were employed by neolithic man to saw through chank-shells and to fashion bangles therefrom. This I cannot accept.

GUJARAT AND KATHIAWAR.

Mr. Bruce Foote's labours prove that the custom of using chank bangles was widely spread and that chank-bangle factories were numerous in these two provinces in ancient times.

The finds which he records are as follows :—

In Kathiawar :—

(a) Damnagar, Amreli Prant. In the fields (presumably upon the surface) north of the camping tope at this town a great number of chank bangles in a fragmentary condition were found and of these 41 pieces are represented in the Museum collection. Three working fragments were also found at the same place, together with a couple of cowries, and a Trochus shell ground upon three sides. Associated were such neoliths as a basalt corncrusher, a bloodstone hammer and chert and agate cores.

(b) Babapur. At this village situated 13 miles westward of Amreli, the alluvium of the left bank of the Shitranji river yielded a large and important series of neolithic chert flakes, scrapers, slingstones, and cores in association with 13 fragments of finished chank bangles, together with two working fragments and part of the columella of a chank. Several of the flint flakes are worked upon one or both edges, and one of the bangle fragments exhibits a chaste design executed with considerable delicacy. The other bangles are of plain and crude design.

(c) Ambavalli. Seventy-one fragments of broken bangles from an old site at this place are represented in the Museum collection (Nos. 3622-1 to 65 and 81 to 89). Of these the greater number are ornamented by pattern grooving and many show an elaboration of design as great as those now manufactured in Bengal. The designs in many instances are precisely the same as those in vogue to-day.

Associated with these bangle fragments were numerous portions of sawn sections of chank-shells, constituting the rough working material required by the bangle carver ; 33 fragments are shown (Nos. 3622-63 to 65 and 90 to 119).

With the exception of a few unimportant potsherds the only other object of importance found at this site was a small iron knife with tang. No stone implements were discovered, and no information is given as to the precise conditions under which any of the exhibits were found ; presumably they lay on the surface of the ground examined.

(d) Sonnaria. Fragments of two chank bangles of simple pattern apparently found on the ground surface. A chert scraper comes from the same locality.

(e) Kodinar. On the surface of Mr. Foote's camping ground were found several sawn portions of chank-shell, two being shoulder slices such as are found in the wastage from a bangle workshop.

(f) Vālābhipur (the modern Walah). From the ruins of the ancient city Mr. Foote obtained a large and most interesting series of chank-bangle fragments, 62 in number. With them were a smaller number (7) of sawn working sections. A few marine shells (*Nerita, Nassa, Ostrea,* and *Conus*) were also found among the ruins.

In Gujarat —

(a) Sigam, on north. bank of the Heran River. Five weathered sawn working sections of chanks are represented in the collection from this site. No finished remains of bangles were seen. The site yielded a variety of neolithic flakes and cores and two sandstone hammers or pestles. No indication is given of the precise mode of occurrence, but I conclude they were all surface finds.

(b) Kamrej, 12 miles north-east of Surat. The summit of a small islet in the Tapti river at this place yielded three sawn shoulder shoes (workshop waste) of chank shells and a single fragment of finished bangle. The latter is remarkable for the peculiarity and elegance of its pattern, a broad and closely worked zig-zag groove such as I have never seen either among ancient bangle fragments or on any of the present-day productions of Bengal. With these chank remains were two fragments of sandstone hammers.

This site is notable as being on an islet in the Tapti river protected against assault by steep and almost inaccessible sides—a place very defensible and therefore an ideal place for the settlement of craftsmen.

(c) Mahuri, in Vijapur Taluk. From " an old site at the head of the gully system which cuts deep into the alluvium of the Sabarmati " at this place, a series of working sections and waste pieces of chank shell was found sufficiently numerous and varied to convince me, after examination of the fragments, that a bangle workshop undoubtedly existed here at a remote period. The presence of sawn waste associated with sawn working sections is conclusive.

Of completed bangles the remains found were few (8 are shown in the collection) but of these, three are of special interest on account of the elaboration of ornament exhibited. Two of these fragments are of broad bangles richly carved in patterns very closely approximate if indeed not identical with forms in use at the present day. The third fine example is a tiny fragment of the narrow form of bangle known as *churi* in Bengal, usually worn in sets of three on each wrist. The other fragments found are of simpler patterns.

An interesting associated find was that of a small " finial " carved out of shell, probably mother-of-pearl. It is identical in form with a mother-of-pearl nose-pendant now in use among the poorer castes in some country districts in Bengal. An example which I purchased in Eastern Bengal is carved from the shell of a river mussel (*Unio* sp.). From the alluvium at Mahuri whence the bangle fragments came, a few neolithic implements, chert flakes and scrapers principally, were unearthed, together with several noticeable pieces of pottery. Of the latter, one is of special importance as it affords some evidence better than the neoliths touching the age of the bangle factory once situated at this place. It is a small headless figure of a sacred bull, of polished earthenware, red externally and black within. Two garlands are indicated around the hump by means of rows of tiny impressed punctures and there can be little doubt that it is of early Brahmanical age.

(a) Kheralu. A single fragment of a sawn working section of chank shell was found on the surface of the loess at this place.

Eight sites can clearly be indicated as probable centres of the chank-bangle industry in Gujarat and Kathiawar, namely :—(a) Sigam, Hiran Valley, Baroda Prant, (b) Kamrej, on the Tapti, (c) Mahuri, on the left bank of the Sabarmati, Baroda State, with (d) Ambavalli, (e) Damnagar, (f) Kodinar, and (g) in and on the alluvium of the Shitranji river above Babapur, all four in Amreli Prant, Kathiawar, also (h) Vālābhipur in Vala State, Kathiawar. At all these places working fragments of chank shells have been found. The most important sites appear to have been those at Mahuri in Gujarat and Ambavalli and Vālābhipur in Kathiawar. The unworked sections and waste pieces of shells found at these three places are so numerous, and so characteristic in their form of stages in shell-bangle manufacture, that we are perforce compelled to admit these sites as having been in old times locations of important factories, a conclusion to which further weight is given by the discovery at each of these places of fragments of completed bangles, in many instances of highly decorated patterns. At Ambavalli and Vālābhipur fragments of finished bangles are especially plentiful ; the ornamentation is well executed and exhibits considerable taste, a high degree of skill, and undoubtedly the employment of effective tools of several sorts—saws, drills, and files. Iron is the only metal suitable for making tools fit for carving the extremely hard substance of chank shells and it is of the greatest interest and significance that at the Ambavalli site associated with the many fragments of worked and unworked chank circlets found there, an iron knife with a tang was discovered which from personal examination I am satisfied may well represent such a chank-saw as is to-day in common use in Bengal chank factories for cutting patterns upon the bangles.

From a consideration of the details given above a certain number of facts and conclusions of importance emerge, to wit :—

(a) In all cases the fragments of bangles and of chank shells appear to have been surface finds. In several cases this is definitely stated and in the remainder wherever no statement of horizon is given, the context points to a like provenance. From this it follows that association with neolithic artifacts in itself has little value or significance ; both neoliths and chank fragments are practically indestructible by atmospheric weathering agencies and their association may merely connote the fact that particular surface areas have suffered little or no denudation or change since neolithic times whereby the broken implements and discarded ornaments of a later age have mingled with those of an earlier one. Or it may be the result of the artifacts of different ages having been weathered out of different alluvial strata in such way that they come eventually to lie together at a lower level of the original ground or else in some newer river deposit into which floods may have rolled them.

(b) The facts already noted that all sections of chank shells, working pieces as well as wastage scraps, show cleanly sawn surfaces as verified by examination of the originals now in the Madras Museum, and that these surfaces show series of striae *often at two or more angles to one another*, are sufficient to negative the tentative suggestion made by Mr. Bruce Foote assigning a neolithic origin to the workmanship. Neither serrate nor biserrate chert flake saws however delicately made could possibly produce such cleanly sawn sections as we see represented in the collection. The aid of thin metal saws must be invoked and it is most significant that in two instances (Ambavalli in Kathiawar and Muski in the Raichur Doab) fragments of iron knives were found associated with the remnants of chank working sections. In several other cases (Srinivaspur in Mysore, Havaligi Hill in Anantapur, and Bastipad in Kurnul) pieces of iron slag were found in association.

As the working sections of chank shells retain visible evidence of being sawn by means of a metal (iron) saw and as iron fragments are frequently associated with them, the evidence is to me satisfactory that the age of the former cannot possibly be neolithic ; knowledge of the manufacture of iron into somewhat elaborate tools—saws, files, and drills—must have been possessed by the bangle makers. This would appear therefore to rule out the early iron age, when iron weapons and tools were of primitive design.

Incidentally this conclusion is likely to affect the estimate of age accorded to the potsherds so frequently associated with fragments of chank bangles and to render doubtful their identification as neolithic or even of early iron age.

(c) Three sites alone give other than negative evidence in regard to age. These are Gudivada in Kistna district, Vālābhipur in Kathiawar and Mahuri in Gujarat. The remains at the first named are indubitably Buddhistic while the occurrence of a figurine of a bull with a double garland round the hump points distinctly to an age when the adherents of Brahmanism were in the land holding in especial reverence Siva's sacred bull. Most important find of all was that made in the ruins of Vālābhipur, for the history of this old city is fairly well known ; the dates of many of the great events that happened there are on record and the descriptions of two Chinese Buddhist pilgrims who visited the city are extant. The story of Vālābhipur goes back some centuries before the Christian era and for long it was the seat of the Valabhis, a Rajput race, and the centre of their rule, till the middle of the eighth century when the last of the line was overthrown by Arab invaders from Sind. Valabhi was visited by the Chinese pilgrim, Hiuen Tsang, in the course of his fifteen years' sojourn in India (A.D. 630-645) and by I. Tsing in the succeeding century. Both pilgrims describe it as a large and flourishing city and a great centre of Buddhist learning, its streets and schools crowded with students. The reigning dynasty, themselves of the Brahman faith, appear to have been tolerant of Buddhism like many of their contemporaries. ˙In Hiuen Tsang's days the latter religion was still followed by great numbers of the populace, especially in Orissa and Southern India ; elsewhere Hinduism was rapidly becoming the popular religion and the mass of the people were of this faith when the last Valabhi dynasty ended.

As the chank is a religious symbol both to Hindus and to Buddhists, we may reasonably conclude that the remains of chank bangles found in Válābhipur were made for the use of the women of the town and neighbourhood not later than the eighth century. The trade must have been long established at that time to judge by the excellence of the work turned out, which fully equals that of average Bengal workmanship of the present day.

. Taking all facts into consideration I am inclined to date the majority of the bangle fragments found in the Kathiawar and Gujarat sites as roughly contemporary with the Válābhipur specimens or at most not antedating them by more than 300 to 400 years.

To date the Deccan chank-bangle factories is more difficult ; one outstanding fact is the simplicity of all the patterns. The great majority are devoid of ornament save for a boss roughly carved at one side. This plainness of design would seem to bespeak less skill on the part of the Deccan workman than on that of his fellow craftsman in Gujarat. If that be the explanation, and if it be not due to lack of taste or of the means to pay for good work on the part of the buyers, then we may reasonably date the majority of these fragments back to the first few centuries before or after the beginning of the Christian era. The presence among the pottery mixed with the bangle fragments found near Srinivaspur in Mysore of a flat sherd similar in pattern to one found with the Buddhist remains at Gudivada in Kistna District is noteworthy as lending further countenance to this conclusion.

(d) The finds made by Mr. Bruce Foote argue two great centres of chank-bangle manufacture and usage apart from that in the extreme south of the Madras Presidency, namely, one in the Southern Deccan and the other round the shores of the Gulf of Cambay. It is most probable that other centres of the industry did exist, but at present there is no direct evidence to this effect. For instance, it is not likely that an industry which was firmly established in Eastern Bengal at the time of the arrival of the Portuguese in India [1] and of Tavernier's travels in the seventeenth century, and which. continues to flourish at the present day, should be of modern growth.

With regard to the third known seat of the industry in ancient times, that which flourished in. the early centuries of the Christian era in the Tinnevelly district, its geographical location in the coastal section of the Pandyan kingdom made it the natural centre and home of a great chank-cutting industry. Its Pandyan sovereigns were from time immemorial overlords of the Pearl and Chank Fisheries of the Gulf of Mannar and

---

[1] Garcia da Orta writing in the sixteenth century states that the chank was then an article of importance in the Bengal trade, *though less valuable than formerly.*

Palk Bay, the most important source of supply of the raw material then and now, and it is a curious vagary of trade that the present seat of the industry should be situated 1,500 miles from the scene of the fishery.

From the fact that among a few widely separated castes, sub-castes and tribes of the extreme south of India, including among others the Kotas of the Nilgiri Hills and certain sections of the Vellalans and Idaiyans in the inland Coimbatore district, the custom prevails of wearing chank bangles for ceremonial reasons, we may also reasonably infer the former wider prevalence of the custom. Indeed it is probable that the custom was at one time prevalent throughout a large section of Southern India.

Kathiawar and adjacent Gujarat are also both maritime provinces and this geographical situation is the key to the location of a chank-bangle industry in those provinces in early times ; the coast of Kathiawar is the only considerable source of chank shells apart from the Gulf of Mannar and Palk Bay. No chank-cutting is now done either in Kathiawar or Gujarat ; the women there have abandoned their former habit of wearing chank bangles and all the shells fished in this locality are exported from Bombay to Bengal where they are known in trade as " Sūrti " shells, Surat having been the port of shipment prior to the rise of Bombay.

Why the Southern Deccan should once have been the home of a shell-cutting industry is not so easy of explanation, seeing that it is situated in the heart of the country and distant from 400 to 500 miles from the nearest sources of supply (Rameswaram and the Tanjore coast). Possibly the location of this trade in the Deccan was due to the superior skill as craftsmen of the people in this region inherited from stone-using ancestors who found in the quartzite and trap rocks of the district more suitable material for their weapons and tools than the men to the southward where intractable gneiss constitutes all the rocky outcrops. Certainly in prehistoric times, Bellary, Kurnul and Cuddapah were more thickly populated than the country to the south if we may judge from the evidence of the number of stone implements found respectively in these two sections of India. The neolithic remains of these Deccan craftsmen show their makers to have been comparatively highly-skilled workers and with the discovery of the use of iron, hæmatite ore being abundant in Bellary, the men of this district may reasonably be supposed to have developed special skill in the working of the new material into tools and in the manufacture of many articles, ornamental as well as useful, with the aid of these improved tools. Add to this the natural conservatism of tribes isolated from the coast by hill ranges—the customs and manners of the Deccan tribes have been less changed by contact and intermixture with surrounding races than the majority of the tribes or races living in the coastal plains. To these inland people the wonder of the great shell honoured by their gods would appeal vividly ; the mystery to them of its origin would confer added importance and, as we find the wild hill tribes of Thibet, Assam, and Bhutan do at the present day, they would end by endowing ornaments made from it with mysterious powers of ensuring well-being and good luck, even as the

Buddhist cartmen of Ceylon and their Hindu brethren throughout the Southern Carnatic adorn their bulls with a chank shell as an amulet against the evil eye.

Chank shells for the Deccan bangle workshops may probably have come from the Tanjore coast, this being the nearest source of supply. The Tanjore fishery appears to have been fairly lucrative down to 1826 when economic changes caused a collapse of the industry. Tirumalavasal at the mouth of one of the northern branches of the Kaveri is the centre of the chank-fishery on this part of the coast and is not far from Kaveri-pattanam, once the chief port of the Chola kingdom and in the height of its prosperity in the early centuries of the Christian era. From Kaveri-pattanam to the inland districts of Kurnul and Bellary the transit of goods would be comparatively easy and safe ; coasters would be used to the mouth of the Kistna, 350 miles to the north, whence river craft would carry the goods direct to their destination, 200 miles inland. Or it may be that the shells required in the industry were fished further south, for we have mention by Cosmas Indico-Pleustes in the sixth century (*circa* 545) of a place called Marallo on the continent adjoining Ceylon, where a shell called by him κοχλίους (Kochlious) [1] was produced in quantity, and Yule in " Cathay and the Way Thither " (London, 1866), Vol. I, p. 81, suggests that this Marallo is the same word as Marawa, the name of the ruling caste in the district of Rāmnād ; if this be accepted, the reference would indicate the chank-fishery carried on off the coast of the Marawar country and now operated by lessees of the Raja of Rāmnād. Again, a chank fishery, the most productive in the world, exists to-day in the shallow seas in the neighbourhood of Jaffna in Ceylon and direct communication by means of large native craft having existed from time immemorial between the north of Ceylon and the port of Masulipatam, for centuries the eastern sea-gate of the Deccan, this fishery may have been drawn upon also to supply the needs of the latter locality.

The cause of the cessation of the chank industry in the Deccan, Gujarat, and Kathiawar is to be looked for in the constant strife which kept India in a welter of blood through the six centuries of Muhammadan dominance in the land. From the days of Mahmoud of Ghazni, the northern and central portions of India in particular were harried by successive waves of fanatic invaders sweeping down through the north-west passes, and from the thirteenth century onwards to the end of the seventeenth the story of India is that of an unceasing contest between Muhammadan and Hindu for power on the part of the former and for existence and religion on that of the latter. Well may certain old Hindu customs have disappeared ; during the worst periods when the intolerance of the conquerors was at its height, their influence was often exerted towards the suppression of Hindu customs and this, combined with the dislocation of trade consequent upon the general insecurity of the country and the frequent recurrence of raids and widespread warfare, may be considered the main reason for the decay of

---

[1] In the Norman-French dialect still spoken in Jersey and the other Channel islands, the common whelk (*Buccinum*), which is the European representative of the Eastern chank, is known as coqueluche !

and beautiful. This chank is an article of trade to Bengal, and used to be worth more than it is now. The large ones, which we call Buzios, go to Bengal and are worked up very beautifully, remaining very smooth and white. For this only a small quantity is used, the rest being wanted for bracelets and other ornaments. It was the custom in Bengal that no person of distinction who was a virgin could be corrupted unless she had bracelets of the chank shell on her arms.[1] After the arrival of the Patans this custom was neglected and the chank became cheaper in consequence. You see here a chess table at your service where you may see the chank when you like."

[1] Another translation renders this sentence rather differently, to wit :—"There was formerly a custom in Bengal that no virgin in honour and esteem could be corrupted unless it were by placing chank bracelets on her arms." Da Orta's statement refers really to the fact that an essential ceremony in a Bengali marriage consists in placing a chank bangle on each of the bride's wrists ; the marriage would not be formally valid if this were to be omitted.

V

THE SCOPE FOR DEVELOPMENT OF A SHELL-BANGLE INDUSTRY IN OKHAMANDAL

AT the present day the Indian shell-bangle industry is limited to Bengal, where in Dacca, Calcutta, Nadia, and other centres it gives lucrative employment to large numbers of skilled artisans. How economically important it is as an art industry will be realised when we learn that the value of the raw material required ranges between 2 and 2½ lakhs of rupees per annum as given by the importers, an amount probably understated so considerably that it is quite safe to put the average actual value at 2½ lakhs. This is the wholesale value ; the shells usually change hands twice before they reach the actual bangle-workers ; in turn the latter sell their products to bangle merchants who distribute them to shopkeepers and peddlers throughout Bengal, Behar, Assam, Bhutan and Thibet. The value increases several times in this process of manufacture and distribution till it is certain that the final retail value of the products of this industry at a conservative estimate is considerably over 15 lakhs of rupees, or £100,000. From one shell an average of three wide bangles can be cut ; the wholesale value of good shells may be taken at from 1 to 2 annas each, whereas the one and a half pair of bangles that one shell produces will be sold at anything from 9 annas to rupees 18 according to the amount of ornamentation. Certainly the three bangles yielded by one shell have an average retail value of not less than one rupee. On this basis the shells increase in value eight times on the import value if we take the latter at the high rate of 2 annas each.

The tools employed by the shell-cutters and gravers are of the simplest and most primitive description ; they consist of heavy double-handled saws employed in sawing the circlets which constitute the working sections for the bangle carver (*vide* Pl. IV, Figs. 1 and 2) and of small handsaws (Pl. V, Figs. 1 and 2), bow-drills and files for smoothing and carving the patterns. No labour-saving devices are used ; the tools with the exception of European-made files, are undoubtedly identical with those used 1,500 years ago, in particular the great saw—the most important of these tools—which is referred to in an ancient Tamil poem quoted on page 59.

The present concentration of the shell-bangle industry in Bengal is consequent

Fig. 1.—Sectioning chank shells in a Dacca workshop.

Fig. 2.—Sharpening a chank saw, Dacca.

Fig. 3.—Breaking away the remains of the septum from a sawn chank circle (working section).

Fig. 4.—Rubbing down the inner surface of a chank working section, Dinajpur, Bengal.

[*Photographed by J. Hornell.*]

Fig. 1.—Carving an incised pattern, Rangpur.

Fig. 3.—Forming a simple ridge pattern by rubbing down the sections on a stone, Rangpur, Bengal.

Fig. 2.—Sharpening an engraving saw, Dinajpur.

Fig. 4.—Rubbing down cinnabar to colour lacquer red, Dinajpur.

[*Photographed by J. Horrell.*]

Fig. 2.—Method of using rest when finishing off an inlaid lacquer bangle, Dinajpur.

Fig. 1.—Lacquering marriage bangles, Dinajpur.

upon this province being the only one in India where shell bangles are extensively used by high and low among the female population. Even in Bengal itself shell-cutting proper—the sawing of the shells into working sections—is restricted to a very few centres ; this part of the work requires a great amount of skill such as can be acquired only after a long and toilsome apprenticeship, begun at an early age ; shell-slicing calls for the possession of a highly trained eye, perfect steadiness of hand and arm and an ironlike capacity to sit for long hours in a position of great discomfort.. As a consequence the sawing of working sections is limited to a few centres and a good cutter is a valuable asset to his employer. To retain a hold upon these men, employers willingly give large advances in cash.

The towns and villages where the working sections are fashioned into the bangle patterns favoured locally are much more numerous than the sawing centres ; the bangle-workers in the great majority of such places do not attempt to slice the shell itself ; they depend upon the wholesale sawyers of Dacca and a few other great centres for their working sections. Particularly is this the case with Muhammadan bangle-workers, for these men are comparatively new recruits to the trade and therefore are employed chiefly in the less skilled sections of the industry and in fashioning bangles of the simple and crude patterns affected by the poorest and most ignorant of the population.

In spite of the two great advantages possessed by the bangle trade in Bengal—(a) location in the centre of the only province where chank-bangle wearing continues to be widely prevalent, and (b) the possession of a caste of highly skilled hereditary chank-bangle sawyers and carvers, there is ample scope for the establishment of a large chank-sawing industry in those localities where the shells are fished extensively—in Kathiawar and on the coast of the Gulf of Mannar. Indeed the proposition is one economically sound. At present the whole shell has to be transported from the place of fishing to Calcutta and Dacca, distances in both cases upwards of 1,200 miles, where the mouth part, the apex, and the columella, constituting over 60 per cent. of the total weight, are cut or broken away and thrown out as waste in the process of manufacture. Under present conditions freight and handling charges have thus to be paid on more than double the weight of the material actually employed. A radical change in the method of preparing working sections would, however, be necessitated as the present hand-sawing of the shell is an art too difficult for unskilled labour to acquire and I believe no Bengal sawyers would leave their own districts to work elsewhere. The one alternative is to adopt a form of machine saw capable of employment by comparatively unskilled workpeople. A band saw working with emery, or preferably with carborundum powder, probably would give satisfactory results, the motive power being either foot power leaving the hands free to hold and to guide the shell or, preferably, a small oil engine or else electric current in towns where the latter is available. The introduction of a chank-cutting industry into Okhamandal is, I believe, eminently feasible and practical and commercially sound. To begin with there is a considerable supply of shells available on the Okhamandal coast itself, and if this were to prove inadequate, it could be supplemented by supplies drawn from other parts of the coastline of Kathiawar.

The supply being therefore assured, the chank-sawing industry might be developed along either of two lines, either the shells might be sawn into working sections and these exported to Bengal in this partly manufactured state, or the finished bangle might be produced. With provision of a power machine suitable for cutting sections, this could readily be adapted to do smoothing, carving, and polishing, so that an intelligent workman —especially if he be one having some knowledge of lapidary work—if he were given a series of simple patterns to begin with and direction how to use the machine he has to handle, should have no insuperable difficulty in doing passable work in the course of a few weeks. Once he acquires a command over his machine and the ability to carve simple patterns, the power to attack more complicated patterns—such as those depicted on Plate VII—would soon come. As to a market, Bengal has now increased so greatly in population that the chank-bangle workers there begin to find difficulty in obtaining a sufficient supply of the raw material, so even if no local demand should develop in Okhamandal, the Bengal market would absorb readily all that an Okhamandal workshop would be likely to produce for years to come. It seems, extremely probable, however, that a more remunerative local demand may be counted upon to develop concurrently with the progress of the industry. Many thousands of pilgrims resort to the holy shrines of Bēt and Dwarka annually and these people, who value the chank shell as one of the emblems both of Vishnu and Krishna, as a vessel for use in their household religious ceremonies and, in many cases, as a curious souvenir of a memorable pilgrimage, would, I believe, esteem still more a handsome ornament made from this shell for their womenfolk, particularly if judicious efforts were made to draw attention to the appropriateness of such an ornament on the hands of Hindu ladies ; the beauty of its pattern, its snow-white colour—emblem of purity—and its general superiority to the gaudy glass bangles of European manufacture that not infrequently inflict serious wounds on the wrist when they are accidentally broken. Patriotic preference for a home-made ornament, especially when backed up by the more solid advantages of greater real beauty and greater strength and lasting property, should also count for much. Lastly, a most valuable potential market for high-priced chank bangles certainly exists in Bombay, Karachi, and other great cities where European and American tourists are to be found in yearly increasing numbers. These people are all eager to take home some " curiosities " of the East, preferably such as are not too bulky ; from my own knowledge I can say that well-carved chank bangles do appeal greatly to these visitors from abroad and it is only lack of enterprise on the part of Bengal chank bangle manufacturers that has prevented the exploitation of this promising field. Time after time people who have seen the chank bangles in my collection have expressed their desire to purchase—a desire baulked by the fact that none are to be had in any shops visited by tourists in any city in India.

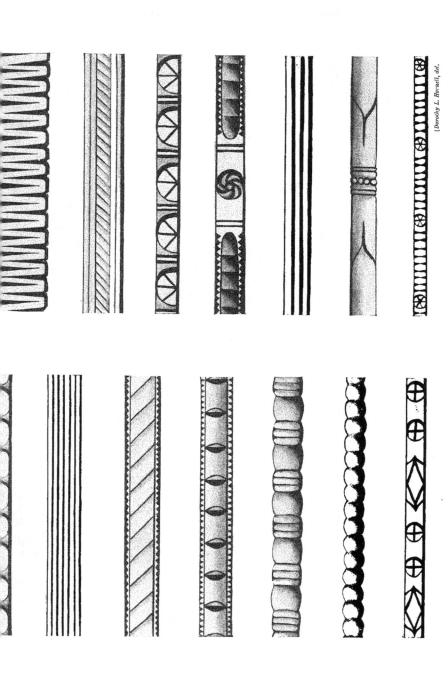

employed in inlaying ; Garcia da Orta mentions the use of squares of chank shell in making the white squares of chess-boards.

EXPLANATION OF PLATES.

PLATE I (FRONTISPIECE).

Figs. 1, 2, and 3.—Sinistral chanks respectively in the Satya Bhamaji, Shank Narayan, and Lakshmi Temples, Bēt, Kathiawar.

PLATE II.

Figs. 1 and 2.—Immature chanks (*Turbinella pyrum*) from Okhamandal; both show persistence of the protoconch.

Fig 3.—Adult chank from Okhamandal.

Fig. 4.—Elongated variety from the Andaman Islands. Also two very young individuals with protoconch well shown.

PLATE III.

Fig. 1.—Group of Cheruman Women wearing necklaces of so-called chank-rings.

Fig. 2.—Under a banyan tree. Selling chank shells to pilgrims returning from Bēt.

PLATE IV.

Fig. 1.—Sectioning chank shells in a Dacca workshop.

Fig. 2.—Sharpening a chank saw, Dacca.

Fig. 3.—Breaking away the remains of the septum from a sawn chank circle (working section).

Fig. 4.—Rubbing down the inner surface of a working section, Dinajpur, Bengal.

PLATE V.

Fig. 1.—Carving an incised pattern, Rangpur.

Fig. 2.—Sharpening an engraving saw, Dinajpur.

Fig. 3.—Forming a simple ridge pattern by rubbing down the sections on a stone, Rangpur, Bengal.

Fig. 4.—Rubbing down Cinnabar (Hingol) to colour lacquer red, Dinajpur.

PLATE VI.

Fig. 1.—Lacquering marriage bangles, Dinajpur.

Fig. 2.—Rest used when finishing off an inlaid lacquer pattern, Dinajpur.

Fig. 3.—Chank-shell waste from ancient bangle factory sites at Korkai, Kayal and Tuticorin (two upper rows) compared with modern waste pieces from Dacca (bottom row, where a working section is also shown).

Fig. 4.—Making children's feeding spouts from chank shells, Karimanal, near Pulicat (Madras).

PLATE VII.

Pattern reconstruction of some of the ancient bangles in the Foote Collection, Madras Museum.

REFERENCES.

The two most frequently quoted works in this Report being :—

THURSTON, EDGAR.—' The Castes and Tribes of South India," 7 Vols., Madras, 1909, and
RISLEY, H. H.—"The Tribes and Castes of Bengal," 2 Vols., 1891,

references thereto are abbreviated to the author's name followed by the number of the volume and the page therein quoted, *e.g.*, Thurston, II. p. 21.

CALCAREOUS SPONGES.

REPORT

ON THE

CALCAREOUS SPONGES

COLLECTED BY

MR. JAMES HORNELL

AT

OKHAMANDAL IN KATTIAWAR IN 1905-6.

BY

ARTHUR DENDY, D.Sc., F.R.S.,

Professor of Zoology in the University of London (King's College).

[With Two Plates.]

THE collection of sponges from Okhamandal placed in my hands by Mr. James Hornell contains six species of Calcarea, unfortunately not in a very good state of preservation. I have found it necessary to describe as new one species of Sycon (*S. grantioides*) and one of Leucandra (*L. dwarkaensis*), together with a variety of *Leucandra donnani* (var. *tenuiradiata*). More interesting than these new forms, however, are the specimens of *Grantessa hastifera* (Row) and *Heteropia glomerosa* (Bowerbank), which enable me to add a good deal to our knowledge of these little known species.

The classification adopted is that of Dendy and Row's [1913] "Classification and Phylogeny of the Calcareous Sponges, with a Reference List of all the described Species, systematically arranged." I have followed my usual practice of quoting the Register Number (R.N.) of each specimen in order to facilitate accurate reference.

1. **Sycon grantioides**, n. sp.—(Plate I, Fig. 1.)

The single specimen, unfortunately in a bad state of preservation, has the form of a slightly compressed cylinder, rather wider above than below, but contracting suddenly to a moderately wide vent (Fig. 1). There is now no peristome, but it is impossible to be certain that there were no hair-like oxea surrounding the vent in life. The dermal surface is minutely reticulate, without conspicuously projecting oxea in its present condition, but this is because the outer ends of the numerous large oxea have all been broken off short. Size of specimen about 16 mm. high by 7 mm. in greatest width. Texture soft and flabby. Colour in spirit white.

The gastral cavity is wide, and the total thickness of the sponge wall only about 1·0 mm. The gastral surface is smooth and pierced by the numerous small apertures of the exhalant canals. The dermal surface is formed by the fusion of the conical outer ends of the radial chambers to form a reticulate pattern. The gastral cortex is moderately thick and pierced by the short exhalant canals. The radial chambers taper towards their distal extremities, while proximally they open, usually in groups, into the short exhalant canals that pierce the gastral cortex. They are provided with well developed diaphragms at the apopyles. They exhibit the " linked " arrangement described by Jenkin [1908B] in his genus Tenthrenodes. The inhalant canals open on the dermal surface by wide apertures between the distal conuli.

The skeleton is typically syconoid in its arrangement, except that the usual tufts of oxea which crown the distal ends of the chambers in typical species of the genus are replaced by large oxea which cannot really be said to be arranged in tufts at all, but extend sometimes through the entire thickness of the sponge-wall, between the chambers, with their distal ends projecting from the surface more or less at right angles. The articulate tubar skeleton consists of many joints, but there are no specially differentiated subgastral sagittal radiates. The spicules are so much broken and eroded that I have not found it practicable to get perfect specimens suitable for figuring. The following descriptions, however, may be taken as substantially correct.

(1) Sagittal triradiates of the many-jointed articulate tubar skeleton ; with very long and slender rays ; the oral angle wider than the paired angles ; the oral rays curved around the chamber as usual and sometimes slightly and irregularly bent. Dimensions of a specimen whose oral rays lie in about the middle of the chamber wall :—Basal ray about 0·2 by 0·005 mm. [1] ; oral rays about 0·12 by 0·005 mm.

(2) Radiates of the gastral skeleton ; usually triradiate but sometimes with a feebly developed apical ray. With wide oral angle and very long and slender facial rays. These spicules are not definitely oriented but form a confused felt-work in the gastral cortex, in which it is very difficult to follow the individual rays for their entire length. I have measured the basal ray up to 0·4 mm. in length, with an oral ray of about 0·2 mm., each having a diameter at the base of about 0·01 mm., but they are usually rather more slender.

---

[1] The first measurement is the length and the second the greatest thickness in all cases except where otherwise stated.

(3) Oxea ; straight or nearly so, and of nearly uniform diameter, but gradually sharp-pointed at their inner ends ; slender in proportion to their great length. The outer ends are all broken off so that I cannot say what they are like. I have measured the remaining portion of the spicule up to about 1·7 by 0·025 mm. This species is not a typical Sycon. In the arrangement of its oxea it much more closely approaches the genus Grantia, from which, however, it must be excluded on account of the absence of a dermal cortex. In the " linked " arrangement of the radial chambers it resembles those species included by Jenkin [1908B] in his genus Tenthrenodes, but that genus cannot be maintained [Dendy and Row 1913].

Register Number and Locality. III. 4. Off Dwarka.

2. **Grantessa hastifera** (Row).—(Plate I, Figs. 2, 2a ; Plate II, Figs. 7a–7f".)
Grantilla hastifera Row [1909].
Grantessa hastifera Dendy [1913].

It is a curious coincidence that this species, first described by my colleague Mr. Row in 1909, and re-described by myself in 1913, in both cases from very inadequate material, should again occur in the present collection. Mr. Hornell's material, however, enables me to add some valuable particulars, especially with regard to the extremely variable external form and the structure of the very remarkable hastate oxea.

The sponge (Figs. 2, 2a) may be described as consisting of thin lamellæ, folded into irregularly tubular or cup-shaped forms ; sometimes forming irregularly proliferating masses of larger and smaller tubes (Fig. 2a). The lamella or sponge-wall is about 1·5 mm. in thickness, but the diameter of the tubes or cups varies from about 2 mm. to at least 22 mm. Unfortunately the sponge is very fragile and the material has been much broken up, so that it is difficult to say anything about the oscula, but these appear to be naked and terminal. The larger fragments (Fig. 2) look like thin, concave, irregular shells, but they probably formed parts of cups in life. All the pieces are possibly parts of the same colony, and it is certain that the wide cups give off narrow cylindrical tubes.

The outer surface is for the most part smooth and subglabrous, but here and there with a few conspicuously projecting spicules. The inner surface (Fig. 2) is pitted·by the numerous openings of the short exhalant canals, arranged in groups. The colour in spirit is dirty white.

The canal system is syconoid, but the material is so badly preserved that it is impossible to make out any details.

The gastral and dermal cortex are each about 0·14 mm. in thickness. The gastral cortical skeleton is made up of the slender rays of tangential gastral triradiates (Fig. 7b) and to a slight extent of the much stouter oral rays of subgastral sagittal triradiates

F

(Fig. 7c). The dermal cortical skeleton is made up of the dermal triradiates (Fig. 7a) arranged tangentially, and of the outer rays of subdermal pseudosagittal triradiates (Fig. 7e). The tubar skeleton approaches but does not completely realise the inarticulate type, the stout subgastral sagittal triradiates being succeeded centrifugally by two or three other sagittal triradiates very similar to themselves (Fig. 7d). The subdermal pseudosagittal triradiates (Fig. 7e) are much slenderer than the subgastral spicules (Fig. 7c), and their centripetally directed rays are commonly grouped in bundles which meet and overlap the opposing basal rays of subgastral or tubar sagittals. Around the oscular margin there is a narrow band of strongly alate triradiates which occasionally develop a short apical ray. There appears to be no oscular fringe of oxea, but a few stout oxea (Figs. 7f, 7f', 7f'') are to be found penetrating the sponge wall more or less at right angles, and often projecting from the dermal surface.

The spicules may be grouped under the following heads :—

1. Dermal triradiates (Fig. 7a). Nearly equiangular but commonly inequiradiate, with indications of a sagittal character. There is no definite orientation, but what appear to be the oral rays are sometimes more or less crooked while the basal ray is straight. All rays gradually and sharply pointed. Size very variable. In a typical example the basal ray measured about 0·42 by 0·025 mm., and the orals 0·36 by 0·025 mm. and 0·3 by 0·025 mm. respectively.

2. Gastral triradiates (Fig. 7b). Very similar to the dermal triradiates but considerably smaller. In a typical example the apparently basal ray, which was straight, measured about 0·16 by 0·02 mm., and the orals, which were very slightly crooked, each about 0·24 by 0·015 mm. Towards the osculum these spicules become very strongly sagittal and exhibit a definite orientation, the oral rays being extended in line with one another and parallel to the oscular margin, while a short apical ray is occasionally developed.

3. Subgastral sagittal triradiates (Fig. 7c). Stout, with oral rays slightly recurved towards the basal, which latter is straight. Oral rays bent towards one another in a plane parallel to the gastral surface, so as to give rise to a deceptive appearance of inequality according to the point of view.[1] All rays gradually sharp-pointed. In a typical example the basal ray measured 0·51 by 0·04 mm., the orals *apparently* 0·37 by 0·035 and 0·25 by 0·035 mm. respectively.

4. Tubar triradiates (Fig. 7d). These are not really distinguishable from the subgastral sagittal triradiates except by their more distal position. The most distally situated show less curvature of the oral rays, and are of smaller size, but they are connected by intermediate forms with the typical subgastral sagittals.

5. Subdermal pseudosagittal [2] triradiates (Fig. 7e). Conspicuously smaller and

[1] Actual inequality may, however, occur to a considerable extent.
[2] For a discussion as to the nature of these spicules *vide infra*, p. 86.

especially more slender than the subgastral sagittals, with the centripetal ray straight and the outer rays more or less bent and asymmetrical. All rays gradually sharp-pointed. In a typical example the centripetal ray measured 0·36 by 0·025 mm., and the outer rays *apparently* 0·24 by 0·025 and 0·17 by 0·025 mm. respectively.

6. Large oxea (Figs. 7*f*, 7*f'*, 7*f''*). These spicules, though not very numerous, are highly characteristic. Their outer ends usually project more or less at right angles from the surface, and often for a great proportion of the length of the spicule, but except over protected areas of the surface they are almost invariably broken off. The form of these spicules affords by far the most characteristic feature of the species, and after a careful study of isolated specimens I am able to add certain particulars under this heading. The entire spicule is slightly curved (Fig. 7*f'*). The inner end is simply sharp-pointed. The outer end is sharp-pointed and flattened like a spear-head, with a sharp knife-edge on either side. On one side (the concave side of the spicule) this knife-edge is simply rounded, on the other it is produced backwards, where it meets the cylindrical shaft of the spicule, into two short, conical teeth. The presence of two teeth can only be clearly seen when the spicule is examined edge on (Figs. 7*f*, 7*f''*), and hence only a single tooth, as seen in side view (Fig. 7*f'*), has hitherto been described. These spicules measure about 0·83 mm. in total length by 0·035 mm. in diameter in the middle, while the spear-head measures about 0·1 mm. in length.

7. Trichoxea ; long and very slender, typically arranged more or less at right angles to the dermal surface, but so scarce that they can hardly be regarded as an essential constituent of the spiculation.

Previously known Distribution. Red Sea (Row) ; Providence I., Indian Ocean (Dendy).

Register Number and Locality. III. 2. Off Dwarka.

3. **Heteropia glomerosa** (Bowerbank).—(Plate I., Figs. 3, 3*a*, 3*b* ; Plate II., Figs. 8*a*–8*g*)

Leuconia glomerosa Bowerbank [1873].

This well-characterised and very beautiful species was first described by Bowerbank in 1873, from dry material collected at Port Elizabeth, South Africa. The type specimen, which is now in the Natural History Department of the British Museum, is a good deal worn, and possibly beach-rolled, and Bowerbank's figure cannot be taken to represent its natural appearance very accurately. The species has not been recorded since its original publication, but it occurs in considerable quantity in Mr. Hornell's collection. Although the locality is so widely separated from that where it was originally found, there can be no doubt as to the identification, which is based upon a careful re-examination of Bowerbank's material. His description and figures are quite insufficient, and in several respects misleading. Thus he says that " the terminal orifices are rarely ciliated ; but when they are so furnished the ciliary fringe is composed of a prolongation of the layer of large acerate spicula." This really applies only to the openings accidentally produced by breaking across of branches, the true oscula are provided with a special fringe of slender, hair-like oxea. Again, he says that there are no " defensive spicula projected from the oscular surface," by which he evidently means no gastral quadriradiates ; such spicules, however, occur in his specimen. The statement that the apices of the subdermal and subgastral triradiates are cemented together where they meet by " keratode " hardly deserves contradiction, the " keratode " being, of course, simply the dried remains of the soft tissues.

Under the circumstances it seems desirable to give a completely new description, based upon Mr. Hornell's spirit-preserved material. I may say, however, that I cannot find any character in which his specimens differ from the type.

The sponge colony (Figs. 3, 3a, 3b) consists of very numerous, rather slender, cylindrical branches, for the most part ascending vertically and lying close together. The branching is very irregular and takes place by the formation of lateral buds at varying levels. Each branch terminates, when fully developed, in a distinct osculum, but the younger buds are blind. The oscula appear to be naked, but are in reality provided with an inconspicuous fringe of slender, hair-like oxea. The projecting portions of these spicules appear always to be broken off. Just within the osculum is a transverse membranous sphincter. All the colonies have evidently been attached below in life, and in one case a portion of the substratum is still present in the form of a barnacle shell.

The largest colony, or piece of a colony, in the collection measures about 32 mm. in height by 38 mm. in greatest breadth, and is composed of about 50 branches measuring up to about 19 mm. in length and 2·5 mm. in diameter. There is considerable variation, both as regards the length of the branches and the compactness of the colony, in different specimens. The surface of the branches is longitudinally striated (Ute-like) owing to the presence of the large dermal oxea, but otherwise smooth. The colour in life was white, in spirit it is dirty white.

The canal system is typically syconoid, the thimble-shaped radial chambers extending at right angles through the wall of the sponge from gastral to dermal cortex.

The preservation of the material is not good enough to enable me to make any detailed histological observations, but the nuclei of the collared cells are apical.

The gastral cortex is fairly thick, and its skeleton is made up of the facial rays of gastral sagittal triradiates and quadriradiates, and the oral rays of subgastral sagittal triradiates. The dermal cortical skeleton is very strongly developed and made up of dermal triradiates, the outer rays of subdermal pseudosagittal triradiates, and huge longitudinally placed oxea, the latter lying for the most part on the inner side of the layer of triradiates. The tubar skeleton is of the inarticulate type (Figs. 8d, 8e), composed of the centrifugal rays of the subgastral triradiates and the centripetal rays of the subdermal triradiates (Fig. 8e, o^1). The former usually extend outwards right through the chamber layer, while the latter may extend inwards as far as the gastral cortex or may fall considerably short of this. Just within the osculum the stout oral rays of the sagittal gastral triradiates, here distinctly alate, are extended parallel with the oscular margin and packed close together in a dense feltwork. Immediately on the *inner* side of this feltwork lie the hair-like oxea of the oscular fringe. The oscular fringe thus arises from the gastral cortex and the colossal dermal oxea take no part in its formation.

The spicules may be grouped under the following heads :—

1. Dermal triradiates (Fig. 8a). These are nearly equiangular but sagittal owing to the greater length of the basal ray. The rays are usually straight and rather slender, tapering gradually to sharp points, the basal ray being a good deal longer and rather more slender than the orals. In a typical example the oral rays measured about 0·11 by 0·01 mm. and the basal ʼabout 0·24 by 0·008 mm. There is a good deal of irregularity in the arrangement of these spicules, but typically the basal ray points away from the osculum as usual.

2. Gastral triradiates (Fig. 8b). These are a good deal larger than the dermal triradiates and more markedly sagittal, the basal ray being usually very long and slender, while the oral rays are often more or less curved backwards. The oral rays are sometimes very unequally developed, one being much longer than the other. The rays are slender and more or less gradually sharp-pointed. In a typical example the oral rays measured 0·2 by 0·012 mm. and 0·17 by 0·012 mm. respectively, and the basal ray 0·44 by 0·012 mm. The basal ray is typically directed away from the osculum as usual. Near the osculum the oral rays are curved backwards so much as to lie almost in a line with one another, and are much more strongly developed than the basal.

3. Gastral quadriradiates (Fig. 8c). These resemble the gastral triradiates, but they are decidedly stouter. The apical ray may be very strongly developed (Fig. 8c, a.r.), straight and sharply pointed, and directed obliquely upwards in the gastral cavity. These spicules are not very numerous.

4. Subgastral sagittal triradiates (Fig. 8d). These are typical alate spicules, with slender, gradually sharp-pointed rays. The recurved oral rays are extended almost in line with one another in the deeper part of the gastral cortex; the straight basal ray extends through the chamber layer to the dermal cortex. In a typical example the oral rays measured about 0·145 by 0·01 mm. ; the basal about 0·245 by 0·01 mm.

5. Subdermal pseudosagittal triradiates (Figs. 8e, 8f). The form and arrangement of these spicules strongly support the view[1] that they are really distal
tubar triradiates which have undergone rotation so that the original basal ray (b)
has come to lie in or below the dermal cortex, while one of the original oral
rays (o^1) has come to be directed inwards and has become more or less elongated.
I had previously looked upon this inwardly pointing ray as the basal ray of a
sagittal spicule, and upon the true basal ray as one of the orals. It will be well
in future to speak of the former (Figs. 8e, 8f, o^1) simply as the centripetal ray,
and the others as the outer or dermal rays. In the present case the dermal rays are
asymmetrical and it is quite easy to see which is the original basal ray (b), for
it is straight, while the other (o^2) is often more or less curved or crooked, and
really forms a pair with the centripetal ray, which is also frequently bent. The
centripetal rays usually lie in close juxtaposition with the centrifugal rays of the
subgastral sagittal triradiates. In a typical example the centripetal ray measured
about 0·13 by 0·01 mm., and the dermal rays about 0·11 by 0·01 mm., but the
centripetal ray may be more elongated. All the rays are more or less gradually
sharp-pointed.

6. Colossal oxea of the dermal cortex (Fig. 8g). These spicules are fusiform
but commonly thicker at one end than at the other. They vary much in size,
up to about 2·85 by 0·075 mm.

7. Hair-like oxea of the oscular fringe. Straight and very slender, only about
0·004 mm. in maximum diameter. Their length is probably about 0·6 mm., but
they are almost invariably broken short in preparations.

Previously known Distribution. Port Elizabeth, South Africa (Bowerbank).

Register Numbers and Localities. I., a number of good colonies from Okhamandal
Point (off Buoy ; 5. 1. 06) ; IV. 6, two small fragments from the S.W. Coast of Beyt
Island.

4. Leucandra donnani Dendy, var. tenuiradiata nov. (Plate I., Figs. 4, 4a, 4b ; Plate II., Figs. 9a–9d).

Leucandra donnani Dendy [1905].

There are five specimens in the collection which I think must certainly be referred
to this species, although the curious differences in the proportions of the spicules
make it desirable to give them a special varietal name.

The external form in the best example (R.N. IV. 8, Fig. 4) agrees closely with
that of the type of the species from Ceylon, even in the curiously curved character
of the whole sponge. The colour, however, is white, owing to the absence of
the pigment granules found in the type. The brittle texture is pronounced and
has resulted in the breaking of the specimen in two.

[1] *Cf.* Dendy and Row [1913], p. 750.

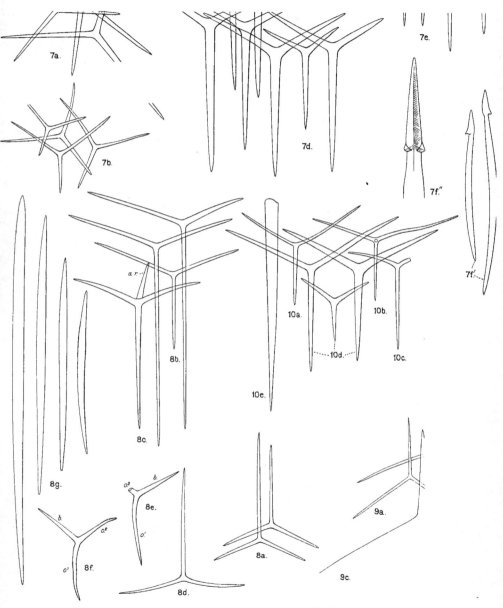

7a.

7b.

7d.

7e.

7f″.

7f′.

8b.

a r

10a.

10b.

10d.

10c.

10e.

8c.

8g.

8e.

b

o²

8f.

b

o²

o³

o′

8a.

8d.

9a.

9c.

CALCAREOUS SPONGES.

The skeleton is composed, as in the type, of a thin layer of slender dermal triradiates (Fig. 9a), a thin layer of slender gastral quadriradiates (Fig. 9b), with some triradiates, and a thick layer of much larger triradiates (Fig. 9c) in the chamber layer between the two. The latter are quite irregularly arranged. The dermal and chamber-layer triradiates are perhaps less strongly sagittal than in the type. There are a few large oxea (Fig. 9d) arranged more or less at right angles to the surface, but with their outer ends broken off short.

The dermal triradiates (Fig. 9a) are considerably smaller and more slender than in the type. The gastral quadriradiates (Fig. 9b) have more slender rays than those of the type, but they may be longer. The chamber-layer triradiates (Fig. 9c) are altogether smaller and of more slender build. The oxea (Fig. 9d), however, are considerably larger than in the type.

This specimen (R.N. IV. 8) may be taken as the type of the variety.

Another specimen (R.N. IV. 17a, Fig. 4a) is of very irregular, sac-like form, and has a delicate oscular fringe of hair-like oxea. There was no such fringe in the type of the species, nor have I been able to find one in R.N. IV. 8, but I cannot say whether or not its absence is due to abrasion. It may possibly be a characteristic feature of the variety. R.N. IV. 17b is a damaged specimen closely resembling R.N. IV. 17a. I have been unable to determine whether or not it has an oscular fringe. There are indications in R.N. IV. 17a that the outer ends of the large oxea may be lance-headed.

R.N. III. 5 (Fig. 4b) is a small specimen, also of curiously curved form, attached to a stony Polyzoon colony (Retepora ?) whereby its surface has been protected from abrasion. The large oxea in this case project far beyond the dermal surface, and many, if not all, of them are distinctly lance-headed. There is also a feebly developed oscular fringe of trichoxea.

The chief distinguishing features of the variety appear to be the comparatively large size of the oxea and the comparatively small size of the triradiates of the chamber layer. The lance-headed form of the oxea is perhaps a specific character, which I have reason to believe occurs in the type of the species (in which it is shown by the broken off outer end of one spicule very clearly).

Register Numbers and Locality. III. 5, off Dwarka ; IV. 8, IV. 9c, IV. 17a and b, off S.W. Coast of Beyt Island.

5. Leucandra wasinensis (Jenkin). (Plate I. Fig. 5.)

Leucilla wasinensis Jenkin [1908].

Leucandra wasinensis Dendy [1913].

I refer to this species a single small specimen (Fig. 5), about 4 mm. in height (exclusive of the oscular fringe of slender oxea, which is very prominent). The

sponge is a typical leucon person of ovoid form.   The canal system is typical.
The spiculation consists of the following elements :—

(1) Rather large dermal radiates, mostly with a well-developed centripetal apical ray.

(2) Large, irregularly scattered radiates of the chamber layer, mostly, at any rate, without apical rays.

(3) Slender subgastral sagittal radiates, mostly, if not all, with short apical rays ; of typical form and arrangement. These are not mentioned by Jenkin in the case of the type, but probably occur there. I also overlooked them in the specimen recorded by me [1913] from Saya de Malha, but I find on re-examination that they undoubtedly occur therein.

(4) Gastral quadriradiates ; a good deal smaller than the dermal radiates and with well developed, sharp-pointed, curved apical rays projecting into the gastral cavity.

(5) Quadriradiates of the larger exhalant canals ; perhaps merely radiates of the chamber layer with short, thorn-like apical rays.

(6) Stout fusiform oxea projecting radially from the dermal surface ; their outer ends too much broken and corroded for description.

(7) Slender hair-like oxea (trichoxea) of the peristomial fringe.

(8) ? A few rather large microxea at the surface in contact with foreign objects.

A direct comparison with the type has convinced me, as in the case of the Saya de Malha specimen, that there is not sufficient difference to justify a specific separation, especially as there are only single specimens known from the three localities.

Previously known Distribution. Wasin, East Africa (Jenkin) ; Saya de Malha, Indian Ocean (Dendy).

Register Number and Locality. XXXV. 9. Off Dwarka.

6. **Leucandra dwarkaensis,** n.sp. (Plate I. Fig. 6 ; Plate II. Figs. 10a–10e).

The single specimen (Fig. 6) is massive and irregular below, compressed above, and terminating in a wide, slit-like vent. The entire body is sac-shaped, with a very wide gastral cavity. Height about 28 mm. ; width in the middle 17 mm. The outer surface is now nearly smooth, but this is doubtless due to the more or less complete erosion of the outer ends of the large oxea. The oscular margin is very thin and shows the remains of a fringe of close-packed, hair-like oxea. A good deal of foreign material is collected on both dermal and gastral surfaces, including numerous small siliceous oxeote spicules. Colour in spirit dirty white.

The canal system is typical. There is a rather thick gelatinous ectosome containing irregular subdermal cavities which lead into the wide inhalant lacunæ. The flagellate chambers are spherical, about 0·087 mm. in diameter and thickly scattered throughout the choanosome. The gastral cortex is thin ; its inner surface nearly smooth and pierced by numerous exhalant openings.

The arrangement of the skeleton offers no peculiarities. The dermal skeleton is composed of slender sagittal triradiates, tangentially arranged, with a few scattered microxea. The gastral skeleton consists of slender sagittal quadriradiates and triradiates. The skeleton of the chamber layer consists of much larger and stouter triradiates, irregularly scattered, and of the inner portions of large oxea whose outer ends are broken off. There are a few slender subgastral sagittal triradiates and quadriradiates in the younger parts of the sponge towards the vent. There is an oscular fringe of trichoxea.

1. Dermal triradiates (Fig. 10a). Sagittal, with long, slender, gradually sharp-pointed rays. The oral rays usually curved, with wide oral angle. Size varying a good deal, rays commonly somewhat unequal in length, averaging say about 0·28 by 0·017 mm. The arrangement of these spicules, except for the fact that they all lie tangentially, is quite irregular.

2. Gastral quadriradiates (Fig. 10b). Sagittal, facial rays straighter and rather more slender than those of the dermal triradiates, but of about the same length and gradually sharp-pointed. Apical ray moderately long, slender, finely pointed, nearly straight. As usual these spicules become more regularly arranged and strongly alate towards the oscular margin, where also the paired rays become much stouter than the basal ray.

3. Gastral triradiates. Similar to the gastral quadriradiates but without the apical ray.

4. Subgastral sagittal radiates (Fig. 10c). Rays long and slender, gradually sharp-pointed. Basal ray much longer than orals, say about 0·32 by 0·015 mm., while the orals are only about 0·19 mm. long. Occasionally with a short apical ray. Arranged as usual with the basal ray directly centrifugally.

5. Triradiates of the chamber larger (Fig. 10d). Approximately regular, with moderately stout, nearly straight, gradually sharp-pointed rays, varying a good deal in actual size and in proportions ; measuring say about 0·35 by 0·03 mm.

6. Large oxea (Fig. 10e). With their inner ends deeply implanted in the chamber layer, or perhaps even projecting into the gastral cavity, and their outer ends projecting obliquely upwards and outwards from the dermal surface. The inner ends, where perfect, are gradually and finely pointed ; the outer ends are all broken off short close to the dermal surface. The perfect spicule must be nearly straight and very long. The portions remaining in the sponge may measure nearly 2 mm. in length, with a thickness of about 0·05 mm. These spicules are very numerous.

7. Hair-like oxea (trichoxea) of the peristome. These are well developed but all broken off short, their remaining portions being densely crowded together.

8. Microxea. These seem to have been fairly numerous in the dermal layer, but, owing perhaps to the fact that the specimen was first preserved in formalin, they have been extensively eroded and perhaps some of them entirely dissolved away. They seem to have been hastately pointed at one end.

The most characteristic feature of this species appears to be the external form, and especially the wide, slit-like vent, but, in the presence of only a single specimen, it is impossible to say how far this may be constant.

The arrangement of the skeleton and the form and size of the principal spicules agree closely with the corresponding features in *Leucandra donnani* var. *tenuiradiata*, but in the latter I have found no microxea and no subgastral sagittal radiates.

Register Number and Locality. XXIII. 6. Off Dwarka, 15–17 fathoms.

LIST OF LITERATURE REFERRED TO.

1873.   Bowerbank, J. S.   "Contributions to a General History of the Spongiadæ," Part IV.   (*Proc. Zool. Soc. Lond.*, 1873, p. 17.)

1905.   Dendy, A.   "Report on the Sponges Collected by Professor Herdman at Ceylon in 1902." (Reports on the Pearl Oyster Fisheries of the Gulf of Manaar, Vol. III., *Royal Society, London.*)

1913.   Dendy, A.   "Report on the Calcareous Sponges collected by the *Sealark* Expedition in the Indian Ocean." (*Trans. Linn. Soc. Lond. Zoology.* Vol. XVI.)

1913.   Dendy, A., and Row, R. W. H.   "The Classification and Phylogeny of the Calcareous Sponges, with a Reference List of all the described Species, systematically arranged." (*Proc. Zool. Soc. Lond.*, Sept., 1913.)

1908 A.   Jenkin, C. F.   "The Calcareous Sponges."   (The Marine Fauna of Zanzibar and British East Africa, from Collections made by Cyril Crossland, M.A., in the Years 1901 and 1902. *Proc. Zool. Soc. Lond.*, 1908).

1908 B.   Jenkin, C. F.   "Porifera Calcarea."   (*National Antarctic Expedition. Natural History.* Vol. IV.)

1909.   Row, R. W. H.   "Report on the Sponges Collected by Mr. Cyril Crossland in 1904–1905. Part I.   Calcarea." (Reports on the Marine Biology of the Sudanese Red Sea. *Journ. Linn. Soc. Lond. Zoology.* Vol. XXXI.)

,, 4b. *Leucandra donnani* Dendy, var. *tenuiradiata* nov. (R.N. III. 5). × 4.

,, 5. *Leucandra wasinensis* (Jenkin) (R.N. XXXV. 9). × 5½.

,, 6. *Leucandra dwarkaensis* n. sp. (R.N. XXIII. 6). × 1¾.

PLATE II.

Figs. 7a–7f″. *Grantessa hastifera* (Row) (R.N. III. 2).

Fig. 7a. Dermal triradiates. × 90.

,, 7b. Gastral triradiates. × 90.

,, 7c. Subgastral sagittal triradiates. × 90.

,, 7d. Tubar triradiates. × 90.

,, 7e. Subdermal pseudosagittal triradiates. × 90.

,, 7f. Large oxeote, front view. × 90.

,, 7f′. do. do. side view. × 90.

,, 7f″. do. do. outer end, front view. × 360.

Figs. 8a–8g. *Heteropia glomerosa* (Bowerbank) (R.N. I.).

Fig. 8a. Dermal triradiates. × 166.

,, 8b. Gastral triradiates. × 166.

,, 8c. Gastral quadriradiate, a. r., apical ray. × 166.

,, 8d. Subgastral sagittal, and 8e, subdermal pseudosagittal triradiate *in situ*; the outer paired (oral) ray of the latter being broken off. b, basal ray; o¹, centripetal paired ray; o², outer paired ray. × 166.

,, 8f. Complete subdermal pseudosagittal triradiate; lettering as before. × 166.

,, 8g. Dermal oxea. × 50.

Figs. 9a–9d. *Leucandra donnani* (Dendy) var. *tenuiradiata* nov. (R. N. IV. 8).

Fig. 9a. Dermal triradiate. × 90.

,, 9b. Gastral quadriradiate. × 90.

,, 9c. Triradiates of the chamber layer. × 90.

,, 9d. Oxeote. × 90.

Figs. 10a–10e. *Leucandra dwarkaensis*, n. sp. (R.N. XXIII. 6).

,, 10a. Dermal triradiate. × 100.

,, 10b. Gastral quadriradiate. × 100.

,, 10c. Subgastral sagittal triradiate (broken). × 100.

,, 10d. Triradiates of the chamber layer. × 100.

,, 10e. Inner end of large oxeote. × 100.

NON-CALCAREOUS SPONGES

COLLECTED BY

MR. JAMES HORNELL

AT

OKHAMANDAL IN KATTIAWAR IN 1905-6·

BY

ARTHUR DENDY, D.Sc., F.R.S.

Professor of Zoology in the University of London (King's College).

[With Four Plates.]

MR. HORNELL'S collection of Non-Calcareous Sponges contains about fifty-eight species, of which I have been able to identify no fewer than forty-two with previously described forms and to describe fifteen as new. Leaving out of account numerous fragments which were not sufficiently well-preserved for identification, the following is a complete list of the species represented. It will be observed that one species (*Higginsia* sp.) has only been generically identified, but the genus is of sufficient interest to make it worth recording.

ORDER TETRAXONIDA.

SUB-ORDER ASTROTETRAXONIDA.

FAMILY STELLETTIDÆ.

1. *Myriastra (Pilochrota) haeckeli* Sollas.
2. *Jaspis reptans* (Dendy).
3. *Asteropus simplex* (Carter).

31. *Ciocalypta dichotoma* n. sp.
32. *Higginsia* sp.

FAMILY DESMACIDONIDÆ.

33. *Esperella plumosa* (Carter).
34. *Desmacidon minor* n. sp.
35. *Iotrochota baculifera* Ridley.
36. *Guitarra indica* n. sp.
37. *Psammochela elegans* n. gen. et sp.
38. *Chondropsis kirkii* (Carter).
39. *Myxilla arenaria* Dendy.
40. *Clathria corallitincta* Dendy.
41. *Clathria spiculosa* Dendy.
42. *Echinodictyum gorgonioides* n. sp.
43. *Raspailia fruticosa* var. *tenuiramosa* Dendy.
44. *Acarnus tortilis* Topsent.
45. *Bubaris radiata* n. sp.

FAMILY SPIRASTRELLIDÆ.

46. *Spirastrella vagabunda* var. *tubulodigitata* Dendy.
47. *Placospongia carinata* (Bowerbank).

FAMILY CLIONIDÆ.

48. *Cliona coronaria* (Carter).

FAMILY SUBERITIDÆ.

49. *Suberites carnosus* (Johnston) var.
50. *Suberites flabellatus* Carter.
51. *Suberites cruciatus* Dendy.
52. *Polymastia gemmipara* n. sp.

ORDER EUCERATOSA.

FAMILY APLYSILLIDÆ.

53. *Megalopastas retiaria* n. sp.
54. *Darwinella australiensis* Carter.

FAMILY SPONGELIIDÆ.

55. *Spongelia fragilis* var. *ramosa* (Schulze).
56. *Spongelia cinerea* (Keller).
57. *Spongelia elegans* (Nardo) var.

FAMILY SPONGIIDÆ.

58. *Hippospongia clathrata* (Carter).

The fact that it has been possible to identify seventy-four per cent. of these species with previously described forms clearly indicates the progress that has been made in recent years in our knowledge of the Sponges of the Indian Ocean, especially when we consider that Mr. Hornell's collection was made in a locality from which, so far as I am aware, no sponges have hitherto been recorded.

As might naturally be expected, a large proportion of the previously known species are identical with more or less well-known Ceylon forms [cf. Dendy 1905]. The commonest species in the collection is *Esperella plumosa* (Carter), which attains a large size and has a very fully developed and very beautiful spiculation. Other common and characteristic Indian Ocean species are *Myriastra haeckeli, Donatia seychellensis, Chondrilla australiensis, Tetilla dactyloidea, Tetilla hirsuta, Phakellia donnani, Auletta lyrata, Iotrochota baculifera, Clathria corallitincta, Clathria spiculosa, Raspailia fruticosa, Spirastrella vagabunda, Placospongia carinata* and *Hippospongia clathrata*.

Of the fifteen new species, *Tetilla pilula, T. barodensis, Guitarra indica, Psammochela elegans* (for which a new genus is proposed), *Polymastia gemmipara* and *Megalopastas retiaria* may be mentioned as exceptionally interesting forms.

The scarcity of true Horny Sponges (Euceratosa) in the collection is remarkable. There is no true bath sponge and, indeed, only one representative of the family Spongiidæ, viz., the common but useless *Hippospongia clathrata*. On the whole, however, the Sponge-Fauna of Okhamandal is undoubtedly a rich one, and I am very glad to have had the opportunity of investigating it.

All the specimens in the collection seem to have come from shallow water, the greatest depth recorded being seventeen fathoms. A considerable number of them were growing upon the large, branching, parchment-like tubes of a polychæte worm,* which appears to be extremely common. It is unfortunate that so many of the specimens were originally preserved in formalin, a medium which is entirely unsuited for sponges and in which they undergo extensive maceration.

As regards the classification employed in this Report it will be observed that certain innovations have been introduced. The Axinellidæ are included in the Haplosceridæ, and the Spirastrellidæ, Clionidæ and Suberitidæ are placed in the Sigmatotetraxonida. I cannot attempt to justify these changes in this place, but must refer the reader to my Reports on the Sponges of the *Sealark* Expedition, now in course of publication, and to future publications by myself and my colleague, Mr. R. W. H. Row, in which we hope to discuss the question of the

---

* I am indebted to Dr. J. H. Ashworth and Professor Fauvel for the information that this worm is a species of the genus Eunice, possibly *Eunice tubifex* Crossland. [A photograph of a forest of the tubes of this worm draped with masses of zoophytes and polyzoa, as seen at extreme low water on the Kiu littoral, Beyt harbour, is reproduced as Plate VI. in "Marine Resources of Okhamandal," in Part I. of this Report.—J. H.]

classification of the Non-Calcarea from the phylogenetic point of view in some detail.

My account of the six species of Calcareous Sponges collected by Mr. Hornell has already been published [Dendy 1915].

1. Myriastra (Pilochrota) haeckeli Sollas.

Pilochrota haeckeli Sollas [1888].
Stelletta haeckeli Lendenfeld [1903].
Pilochrota haeckeli Dendy [1905].

There are in the collection nineteen specimens which I refer to this species. The smallest are approximately spherical and no larger than a pea. The largest is irregular, like a potato; measures about 45 mm. in length by 33 mm. in transverse diameter, and has three vents. They closely resemble the specimens collected by Professor Herdman at Ceylon, but the thin, membranous lip of the vent appears to be devoid of oxeote spicules.

The largest specimen (R.N. IV. 3) shows to a very marked degree the curious abnormality of some of the triænes which I described and figured in the case of the Ceylon material. The reduction of the rays, however, is carried to such an extent in this specimen that in extreme cases the entire spicule is reduced to a perfectly spherical ball of concentrically laminated silica (opal). A precisely similar modification of tetract megascleres occurs in the two known species of the genus Yodomia [Lebwohl 1914, and Dendy 1916].

Previously known Distribution. Philippine Islands (Sollas); Ceylon (Dendy).

Register Numbers, Localities, &c. II. 5 (four small specimens), off Poshetra, Jan. 7, 1906; IV. 3 (two large specimens), IV. 9 a (seven small specimens), off S.W. coast of Beyt Island; XIV. (small fragment), off S.W. of Beyt, 6.1.06; XV. 2 (four specimens, varying greatly in size), three miles W.N.W. of Samiani Light-house, 17 fms., 22.12.05; XX. 9 (one specimen), Adatra.

2. Jaspis reptans (Dendy).

Coppatias reptans Dendy [1905].

This species is represented in the collection by three specimens, of which it is possible that R.N. XX. 4 and R.N. XX. 7 may be parts of the same. The shape is extremely irregular. R.N. XX. 4 is a flattened, cake-like fragment, measuring about 35 by 25 mm., with a maximum thickness of 12 mm., and a rounded margin except where broken. R.N. XX. 7 is a very irregular fragment of about the same size, throwing off irregular digitiform processes. Both have a coarse, firm, harsh consistency and are of a greyish-fawn colour in spirit (having been first preserved in formalin). R.N. XXXIII. 2 b is a massive but flattened specimen measuring about 50 by 50 mm., and completely overgrown by a specimen of *Reniera semifibrosa*. It has now, in alcohol, a slightly pinkish tint, having also been originally preserved in formalin. These specimens differ from the Ceylon type in the less abundant development of pigment. The oxea are a good deal more robust and longer, especially in R.N. XX. 4 and 7 ; R.N. XXXIII. 2 *b* being intermediate in this respect. The asters also appear to be somewhat larger, though still very minute.

Previously known Distribution. Ceylon (Dendy).

Register Numbers and Localities. XX. 4, 7, Adatra ; XXXIII. 2 *b*, Dhed Mora and adjacent rocky ground between Beyt and Aramra, 1 fm., 21.12.05.

3. Asteropus simplex (Carter).

Stellettinopsis simplex Carter [1879, 1886].

Asteropus simplex Sollas [1888].

Asteropus haeckeli Dendy [1905].

Asteropus simplex Hentschel [1909].

Asteropus simplex Dendy, [1916].

This species is evidently an " epipolasid " form from which the triænes have completely disappeared. It was first described by Carter from Fremantle, Australia, and in the same paper that author also recorded it from Hayti. He subsequently recorded it again from Victoria, Australia, where it was collected by Mr. J. B. Wilson.

The Okhamandal specimen forms a mass of sponge cementing together and filling the interstices in an agglomeration of shells (chiefly Siliquaria) and other *débris*. It is now a very light pinkish-grey in colour, having been first preserved in formalin, and in this respect differs from, at any rate, the Victorian and Haytian specimens, which contain large, dark brown pigment-cells in the more superficial part of the sponge. It also differs from Carter's Australian specimens in the more robust character of the oxea, while the Haytian specimen, of which the type slide is in my possession, appears to be to some extent intermediate in this respect. The spiculation of the Okhamandal sponge is as follows :—

(1) Oxea ; stout, curved, fusiform, sharply pointed ; measuring about 1·7 by 0·07 mm. (or even more) when full-sized, but often smaller. The main skeleton is a confused reticulation of these spicules.

(2) Oxyasters ; with small centrum (if any), and rather few (up to about ten) slender rays, which sometimes seem to be very slightly roughened ; total diameter about 0·03 mm. These spicules have only half the diameter of those of the type as given by Carter ; they also seem to be very local in their distribution, so that they may easily be overlooked if only one sample is examined.

(3) Sanidasters (the " sceptrelliform " spicule of Mr. Carter's description). Extremely numerous, especially in the dermal membrane ; about 0·02 mm. long, with slender axis and rather few, moderately long, irregularly arranged, slender spines.

It is interesting to speculate as to whether or not this widely distributed form has arisen polyphyletically by reduction of species of Ancorina in the different localities where it occurs, or whether such reduction has taken place only once in one locality and been followed by extensive migration. In the present state of our knowledge this question cannot, of course, be settled.

Previously known Distribution. West Australia (Carter); Victoria (Carter); Hayti (Carter); South West Australia (Hentschel); Cargados Carajos (Dendy).

Register Number and Locality. V. 2, S. of Chindi Reef, 6–10 fms., 18.12.05.

4. Geodia variospiculosa Thiele.

Geodia variospiculosa Thiele [1898].
Geodia variospiculosa var. *clavigera* Thiele [1898].
Geodia variospiculosa Lendenfeld [1903].
Geodia variospiculosa var. *typica* Lendenfeld [1910].
Geodia variospiculosa var. *intermedia* Lendenfeld [1910].
Geodia .variospiculosa var. *micraster* Lendenfeld [1910].
Geodia variospiculosa var. *aapta* Lebwohl [1914].

Although this species has hitherto been recorded only from Japanese waters, I have no hesitation in referring to it the only specimen of a geodiid sponge in Mr. Hornell's collection, and in support of my identification I furnish the following details.

The sponge is almost spherical, about 14 mm. in diameter, and has evidently been attached to the substratum by one side. The surface is almost smooth, but slightly uneven. No pores or vents are visible under a pocket lens. The colour in alcohol (after formalin) is very light grey.

There is a thin external fur of small, radially disposed oxea, containing also the cladi of protriænes (and possibly anatriænes). This rests upon the cortical layer of sterrasters, which is about 0·26 mm. thick. The main choanosomal skeleton consists of dense radial bundles composed of large oxea and of the shafts of triænes, whose cladi are for the most part extended just beneath the cortical layer of sterrasters.

Spiculation. (1) Long, slender oxea of the choanosome; straight or nearly so, gradually and finely pointed at each end, measuring about 2·0 by 0·028 mm.

(2) Short oxea of the surface fur; fusiform, almost straight, almost stylote, with the narrow inner end slightly rounded off; measuring about 0·2 by 0·0082 mm.

(3) Orthotriænes·; with long, straight shaft tapering gradually to a slender point, and simple, conical cladi; shaft about 1·8 by 0·05 mm., with cladi about 0·24 by 0·034 mm.

(4) Dichotriænes; resembling (3) but with cladi once bifurcate.

(5) Mesoprotriænes; shaft very long and slender, say about 3·7 by 0·017 mm.;

cladi sharp-pointed, approximately equal in length, about 0·07 by 0·0086 mm.; prolongation of shaft sharp-pointed and almost equal in length to cladi; cladi sometimes irregular.

(6) Anatriænes; cladi sharply recurved, sharply pointed; occasionally split at the apex into two almost parallel branches. Dimensions much the same as for protriænes but shaft rather more slender.

(7) Sterrasters; of normal form, with well-marked hilum; elliptical, measuring about 0·08 by 0·065 mm.

(8) Subcortical spherasters, with numerous fairly long, sharp-pointed rays, total diameter about 0·012 mm. Not sharply distinguishable from (9).

(9) Oxyasters; with few or fairly numerous, slender, perhaps faintly roughened rays; total diameter varying up to about 0·05 mm.

(10) Minute chiasters or strongylospherasters of the dermal layer, about 0·006 mm. in diameter.

Previously known Distribution. Japan (Thiele, Lendenfeld, Lebwohl).

Register Number and Locality. XV. 1, three miles W.N.W. of Samiani Lighthouse, 17 fms., 22.12.05.

5. **Donatia seychellensis** (Wright).

Alemo seychellensis Wright [1881].
Tethya seychellensis Sollas [1888].
Tethya seychellensis Keller [1891].
Tethya seychellensis Topsent [1893].
Tethya ingalli (pars) Lindgren [1898].
Tethya seychellensis Kirkpatrick [1900].
Tethya lyncurium var. c. Dendy [1905].
Donatia Ingalli Topsent [1906].
Donatia Ingalli (pars) Hentschel [1909].
Tethya seychellensis Row [1911].
Donatia seychellensis Dendy [1916].

The three specimens in the collection agree so closely with those obtained by the *Sealark* Expedition in the Indian Ocean and dealt with by me in my Report [1916] that it is unnecessary to describe them in this place. Two of them (R.N. XXVI. 8 *a, b*) are well provided with buds and are also remarkable for the presence of immense numbers of oscillatorian algæ in the cortex, which I have rarely seen before in a Donatia. In all the specimens the sex-radiate condition of the tylasters and large oxyasters is very strongly pronounced though by no means constant, and the latter frequently have branching rays.

Previously known Distribution. Seychelles (Wright); Samboangan (Sollas); Flinders Passage, Torres Straits (Sollas); Red Sea (Keller, Row, Topsent); ? South

West Australia (Hentschel); Gulf of Mannar, Praslin Reef, Egmont Reef and Salomon, Indian Ocean (Dendy); Xmas Island (Kirkpatrick).

Register Numbers, Localities, &c. II. 9, off Poshetra, 7 January, '06; XXVI. 8 *a, b,* Adatra Reefs, 25 December, '05·

6. Tuberella aaptos (Schmidt).

(For Literature and Synonymy, *vide* Topsent [1900]).

There is one very typical specimen of this curious sponge in the collection. It is irregularly tuberous, elongated, with an uneven surface beset here and there with small papillæ, some of which have each a small vent. The texture is compact but fairly compressible, the colour in alcohol (after formalin) brown. Length of specimen about 60 mm., greatest breadth 32 mm., greatest thickness 21 mm.

The main skeleton consists of loose bundles of large strongyloxea radiating towards the surface, with scattered spicules between. The dermal skeleton consists of dense brushes of small styli with outwardly directed apices. The large strongyloxea measure about 1·1 by 0·034 mm.; the small styli about 0·26 by 0·0086 mm.

This specimen agrees very closely in all respects with the description and figures given by Topsent, whose views as to the correct generic name and synonymy I accept provisionally. It seems possible, however, that Gray's generic name Aaptos [1867] may have to be revived, and also that Keller's *Tuberella tethyoides* [1880] may, after all, be a distinct species.

Previously known Distribution. Mediterranean (Schmidt, Lendenfeld, Topsent); Gulf of Mexico (Topsent); S.W. Australia (Hentschel); Aru Islands (Hentschel).

Register Number, Locality, &c. VI., S. of Chindi Reef, 6–10 fms., 18.12.05.

7. Chondrilla australiensis Carter.

Chondrilla australiensis Carter [1873].
Chondrilla australiensis Lendenfeld [1886].
Chondrilla australiensis Lindgren [1898].
Chondrilla australiensis Dendy [1905].
Chondrilla australiensis Hentschel [1909 and 1912].
Chondrilla australiensis Dendy [1916].

Several pieces of considerable size, representing one or more flat, spreading crusts, were obtained from Adatra Reef. They appear to have been originally preserved in formalin, but are now in alcohol, and exhibit the usual light brown colour of the species.

Previously known Distribution. Port Jackson, E. Coast of Australia (Carter, Lendenfeld); Sharks Bay, S.W. Australia (Hentschel); Coast of Cochin China (Lindgren); Ceylon, Cargados Carajos, Amirante, Seychelles (Dendy); Aru Islands (Hentschel).

Register Number, Locality, &c. XXVI. 13, Adatra Reef, 25 December '05·

8. Chondrilla agglutinans, n. sp.—(Plate I., Figs. 1a, 1b.)

The sponge occurs as a sort of matrix, holding together and partially envelopin a mass of Siliquaria and other shell-fragments, together with sand and pebbles.   Th surface is glabrous and approximately smooth except where wrinkled by contractio in spirit.   The colour in spirit is chocolate-brown of varying shades.   Oscula small in irregular groups, their margins flush with the general surface.

The cortex is barely 0·1 mm. thick, and is covered by a very distinct cuticle.   Eve in stained paraffin sections I have been unable to detect the inhalant cortical canals so conspicuous in the closely related C. mixta [Schulze 1877].   It is distinctly fibrous an also contains numerous pigment-cells filled with minute brown granules.   Its oute half is densely packed with spherasters arranged in several layers.

The colour of the sponge appears to be due chiefly to the presence in the choanosom of numerous spherical cells of a brown colour, scattered singly and in dense groups Each of these brown cells is about 0·01 mm. in diameter.   Their colour may, howeve possibly be due to staining by colouring matter extracted by alcohol from the pigment cells.   They remind one of the fat-like bodies described by Schulze [1877] as possibl reserve-material in Chondrosia reniformis, and of the similar bodies described by Carte [1887 bis] in his Chondrosia spurca, and of the bodies regarded as possible symbiotic alga by myself [1905] in Hexadella indica, &c.

There are two kinds of aster present, closely resembling those of Chondrilla mixta viz. spherasters and oxyasters.   The former (Fig. 1a) occur chiefly in the outer part o the cortex, but also sparingly in the inner part of the cortex, and still more sparingl in the choanosome.   They have a very large centrum and numerous smooth, sharp conical rays touching each other at their bases ; the total diameter is about 0·028 mm The oxyasters (Fig. 1b) appear to be confined to the choanosome, where they are ver sparingly scattered.   They have a small or indistinguishable centrum, and comparativel few, rather slender, smooth, sharp-pointed, conical rays, say about eight or ten i number.   The total diameter of this spicule is about 0·02 mm.

Register Number, Locality, &c. V. 1, S. of Chindi Reef, 6–10 fms., 18.12.05.

9. Tetilla dactyloidea (Carter).—(Plate II., Figs. 10a–10c.)

Tethya dactyloidea Carter [1869].
Tethya dactyloidea Carter [1872].
Tethya dactyloidea Carter [1887].
Tetilla dactyloidea Sollas [1888].
Tetilla dactyloidea Keller [1891].
Tetilla dactyloidea var. lingua Annandale [1915].

There are five specimens of this sponge in the collection. They are all subcylindrical, provided with a single vent at the upper extremity and a root-tuft of long, silky spicules at the lower (cf. Figs. 10a–10c). The general surface is smooth and porous, the consistence soft and compressible, the colour in alcohol light grey. The smallest specimen measures about 19 mm. in height by 5·5 mm. in diameter ; the largest 38 mm. by 13 mm., in both cases excluding the root-tuft. In one specimen the root-tuft extends downwards for 18 mm. before meeting the mass of sand-grains with which it is still in connection. In the other specimens a similar mass of sand-grains is attached by the root-tuft close to the lower extremity of the sponge.

In one of the larger specimens, which I cut open, the vent forms the terminal aperture of a cylindrical cloacal chamber about 10 mm. in length and 2·5 mm. in diameter, into which numerous larger and smaller exhalant canals open at various levels, the larger ones being continuations of the cloacal cavity deep down into the body of the sponge (Fig. 10a). Just below the point where the larger canals, coming from below, join to form the cloacal cavity, lies the so-called " nucleus," from which the principal fibres of the skeleton radiate, mostly in a downward direction. Curving gently outwards, these fibres break up, over the general surface, into dense surface-brushes, which, however, do not project sufficiently to render the surface hispid to the naked eye. At the lower extremity of the sponge they are continued downwards, outside the sponge-body, to form the root-tuft.

Throughout the greater part of the sponge-body the fibres appear to be composed exclusively of very long and very slender oxea, and they are crossed at various angles by irregularly scattered oxea of similar form, but perhaps shorter.

The dense surface-brushes are composed mainly of long, slender oxea, but mingled with these occur many protriænes with very slender shaft and almost hair-like cladi of unequal length.

The anatriænes appear to be confined to the lower parts of the sponge, where they seem to form the principal constituents of the descending fibres both inside the sponge and in the root-tuft. Their shafts are very long and slender, hair-like, and their cladomes unusually well developed, with sharp, strongly recurved cladi. Protriænes also occur in the root-tuft, but less abundantly than the anatriænes. The microscleres are minute, slender, contort sigmata, of the ordinary Tetilla type, not very abundant.

This interesting and easily recognisable sponge appears to be characteristic of sandy and muddy flats along the shores of the Indian Ocean. Sollas has expressed a doubt whether the specimen from the Mergui Archipelago identified by Carter as belonging to this species is really specifically identical with the type from the S.E. coast of Arabia, but I do not think it at all likely that there is a specific difference. On the other hand, I have myself [1905] described a distinct, but closely-related species (T. limicola) from Ceylon, differing from T. dactyloidea in external form and in the arrangement of the exhalant canal-system. Annandale has recently [1915] described a variety of *T. dactyloidea* (var. *lingua*) from the Chilka Lake, growing in fresh water.

Previously known Distribution. S.E. coast of Arabia, on shallow sandy bottom near shore (Carter) ; ? on the sandy bottom of the Mahim Estuary, off the Island of Bombay (Carter)[1] ; King Island, Mergui Archipelago (Carter) ; Chilka Lake, Bay of Bengal (Annandale).

Register Number and Locality. II. 2 *a–e*. From muddy shore, Balapur, Jan. 1906.

10. Tetilla hirsuta Dendy.

Tetilla hirsuta Dendy [1889].

Cinachyra hirsuta Lendenfeld [1903].

Tetilla hirsuta Dendy [1905].

I identify with this species two specimens (R.N. XIX., 5, 6), neither of which is in a very good state of preservation and neither of which shows the arrangement of the inhalant and exhalant apertures. They agree closely with the type as regards spiculation, but the triænes are very scarce.

R.N. XXI. 5 may also possibly belong to this species. It contains, however, numerous small, scattered oxea, and may possibly be a specimen of *T. poculifera* Dendy [1905]. It is, however, a mere fragment and cannot be safely identified.

Previously known Distribution. Gulf of Mannar and Ceylon coast (Dendy).

Register Number and Locality. R.N. XIX., 5, 6, Vamiani Point, January 5, '06 ; ? R.N. XXI., 5, off Rupan Bander and Kutchegudh, 4–7 fms., 8.12.05.

11. Tetilla pilula n. sp.—(Plate I., Figs. 2a–2c.)

There are three specimens of this fascinating little sponge in the collection, all closely resembling one another in external form and microscopic details of structure. The form is spherical, with a single small vent surrounded by a distinct spicular, membranous collar. The largest specimen (R.N. IV. 9b) measures only about 5 mm. in diameter, the other two only about 3 mm. The surface is minutely conulose, not visibly hispid, and there is no root-tuft. The texture is rather soft and compressible, and the colour in spirit pale greyish-yellow.

The arrangement of the skeleton is very strongly radial. Dense and closely-placed bundles of slender oxea and triænes radiate outwards from a central "nucleus," while the hair-like shafts of the triænes are often collected together in wavy fibres which also diverge from the centre of the sponge. As they approach the surface the spicule-bundles spread out very gradually into surface-brushes (Fig. 2a) composed of oxea, protriænes and anamonænes, the apices of

---

[1] The evidence of the specific identity in this case is insufficient.

the more distally placed oxea and the cladomes of the more distally placed triænes and monænes projecting slightly beyond the surface. The spicule-bundles are crossed at various angles by loosely scattered oxea.

There is no cortex and (in unstained preparations) no visible distinction between ectosome and choanosome.

Spicules. (1) Oxea ; straight, slender, fusiform ; gradually and finely pointed at each end ; measuring about 0·85 by 0·012 mm.

(2) Protriænes (sometimes diænes ?) ; with equal or unequal, slender, straight, sharp-pointed cladi, and very long, slender shaft tapering off into a fine hair. Shaft measured up to 1·1 mm. in length with a thickness of 0·004 mm. (near the cladal end), and cladi about 0·033 by 0·002 mm. Both shaft and cladi are sometimes of hair-like fineness.

(3) Anamonænes (Fig. 2b) ; shaft measured up to 2·5 mm. in length (and then probably broken off) ; of hair-like fineness throughout the greater part of its length, attaining a thickness of 0·008 mm. immediately below the cladome ; the single cladus sharply recurved, gradually and sharply pointed, measuring about 0·025 by 0·008 mm. (at the base) in a well-developed example.

(4) Slender, spirally twisted (contort) sigmata (Fig. 2c), of the usual Tetilla type ; measuring about 0·009 mm. in a straight between extreme points. Very numerous throughout the sponge.

The most characteristic spicules of this species are undoubtedly the ana-monænes, and it appears to me a very noteworthy fact that, although these occur in immense numbers, I have not met with a single anatriæne or anadiæne. The only other known species in which the anatriænes appear to be represented exclusively by anamonænes is, so far as I am aware, *Tetilla pedifera* Sollas [1888].

In *Tetilla pedifera*, however, there are, according to Sollas, no sigmata, nor any other form of microsclere, an unusual feature which serves at once to distinguish it from *T. pilula*. On account of this character Lendenfeld [1903] has included *T. pedifera* in his genus *Tethyopsilla*, but it is evidently very closely related to *T. pilula*, and it may well be doubted whether *Tethyopsilla* is a monophyletic genus.

In *Tetilla coronida* Sollas [1888] anamonænes occur together with anatriænes.

In *Cinachyra hamata* Lendenfeld [1906] anamonænes sometimes occur alone and sometimes associated with a few anatriænes.

Register Numbers and Localities. IV. 9 *b*, dredged off S.W. coast of Beyt Island ; XXXV. 8 *a, b*, dredged off Dwarka, January '06·

12. Tetilla barodensis n. sp.—(Plate I., Figs. 3a–3d).

The single specimen in the collection is approximately spherical and about

13 mm. in diameter. The surface is not strongly hispid, but is covered with an encrustation of sand-grains between the projecting ends of megascleres. Three small mammiform projections, terminally fringed with projecting spicules, presumably indicate vents ; another smaller one, without a terminal fringe of spicules, may bear inhalant pores. Colour internally pale greyish-yellow.

There is a well-developed, dense cortex, about 0·26 mm. thick, full of granular cells and perhaps to some extent fibrous.

The skeleton radiates from a very dense central " nucleus " and consists of well-defined bundles of oxea and triænes, the bundles being separated from one another by fairly wide intervals free from megascleres.

The cladomes of the orthotriænes and anatriænes for the most part lie in the cortex, but some of them project, along with the ends of some of the oxea, into the encrusting layer of sand and foreign spicules. I have seen no protriænes, neither in sections nor in boiled-out preparations, but it is quite possible that their ends lie outside the cortex and have all been broken off.

Spicules. (1) Oxea ; very long, straight, slender, fusiform, tapering very gradually to a fine point at each end, but one end may taper less gradually than the other ; size up to about 2·9 by 0·05 mm.

(2) Orthotriænes (Fig. 3 *a*) ; shaft tapering very gradually to a long-drawn-out, hair-like extremity ; cladi simple but often irregularly bent, normally conical, sharp-pointed, gently recurved. Dimensions of a typical example : shaft 1·6 by 0·03 mm. (thickness just below cladome), cladi about 0·17 by 0·025 mm. In another example, of different proportions, the shaft measured about 1·0 by 0·037 mm. and the cladi 0·26 by 0·033 mm. Sometimes the hair-like portion of the shaft is abbreviated and abruptly truncated.

(3) Anatriænes (Fig. 3 *b*) ; shaft very long and slender, hair-like, measured up to about 2·75 by 0·007 mm. (thickness just below cladome) ; cladi slender, sharp-pointed, about equal, recurved not very abruptly, measuring about 0·06 by 0·006 mm.

(4) Sigmata (Fig. 3 *c*) ; very slender, spirally curved (contort), measuring about 0·012 mm. in a straight line between extreme points ; of the typical tetillid form ; extremely numerous in the choanosome.

(5) Trichodragmata (Fig. 3 *d*) ; bundles of long, slender, hair-like raphides ; the entire bundle may measure up to about 0·13 by 0·028 mm. Abundantly scattered through the choanosome. Usually they contain fewer raphides and are therefore more slender than the specimen measured.

The most characteristic feature of this species is the presence of the trichodragmata, which, so far as I am aware, are known to occur in only one other species of Tetillidæ, viz. *Cinachyra eurystoma* Keller [1891].

Register Number and Locality. XXIII. 8, off Dwarka, 15–17 fms., 12.12.05.

13. Gellius fibulatus (Schmidt) var. microsigma nov.

Reniera fibulata Schmidt [1862].
Reniera fibulifera Carter [1880].
Gellius fibulatus Topsent [1892].
Gellius fibulatus Dendy [1905].
Gellius ridleyi (pars) Hentschel [1912].

Very little is known about the external form of the European type. Schmidt, however, says that it is not easily distinguishable from small specimens of *Reniera alba*. The latter is described (*loc. cit.*) as a shapeless crust between the branches of Clathria and in other situations, with single vents on the summit of short, projecting tubes. This description applies very well to a small specimen in Mr. Hornell's collection. It consists of a thin crust growing over a mass of calcareous *débris* and giving off three short processes of unequal length, the longest measuring about 10 by 3·5 mm. The processes are hollow, and each bears a terminal vent.

The specimen seems to differ from the European form only in the small size of the sigmata. The oxea measure about 0·25 by 0·01 mm., which seems to agree very well with the type, but the sigmata measure only about 0·0164 mm. from bend to bend.

Previously known Distribution of Species. Adriatic (Schmidt) ; North Atlantic (Topsent) ; Gulf of Mannar, Ceylon Seas (Carter, Dendy) : ? other localities in Indian Ocean (see under *Gellius ridleyi*).

Register Number, Locality, &c. XXIII. 7, off Dwarka, 15–17 fms., 12.12.05.

14. Gellius ridleyi Hentschel.

(For possible Synonymy, see Hentschel [1912]).

I refer to this species a number of irregularly massive specimens bearing large vents at the ends of deep, cylindrical oscular tubes. The largest specimen (R.N. XVII. 1) is a clathrous mass of thick, anastomosing branches, for the most part vertical and more or less fused laterally ; each pierced by a wide, cylindrical oscular tube. Colour in spirit (after formalin), light brown. Texture loose and friable, slightly fibrous. The whole mass measures about 70 mm. in greatest breadth and 65 mm. in height, but it has been a good deal damaged. The oscular tubes measure up to 10 mm. in diameter.

The skeleton is a sub-isodictyal reticulation of oxea, with only a slight tendency to collect in fibres. The oxea are gently curved, gradually and sharply pointed at each end, and measure about 0·2 by 0·0096 mm. ; numerous more slender forms occur, presumably young. The sigmata are simply C-shaped, or only slightly contort, slender, and measure about 0·02 mm. from bend to bend.

Hentschel distinguishes his species from the European *Gellius fibulatus* (Schmidt) by the spicular measurements. He states that in the Atlantic–Mediterranean form the oxea measure more than 0·22 mm. and the sigmata more than 0·028 mm. in length, while in the Indian Ocean form the oxea measure less than 0·2 mm. and the sigmata usually less than 0·025 mm.

It appears to me that these differences are too slight to be of any specific value in themselves, especially as I have in my possession (in Mr. Carter's cabinet) a preparation from a specimen from the Devonshire coast (Budleigh Salterton), in which the oxea measure only 0·14 by 0·006 mm., while the sigmata measure about 0·03 mm. from bend to bend.

On the other hand, I think that it may be possible to distinguish an Indian Ocean species characterised by its robust growth and large oscular tubes. For this form I adopt Hentschel's name *ridleyi*, though I must regard it as very doubtful whether all the specimens from the Indian Ocean which have been referred by various authors to *Gellius fibulatus* can now be referred to *Gellius ridleyi*. Mr. Hornell's collection contains, as we have already seen, another *Gellius* which seems to be quite distinct from *G. ridleyi*, and which I regard as a mere variety of *G. fibulatus*, and I consider it highly probable that the specimens which I described from Ceylon in 1905 belong to *G. fibulatus* rather than to *G. ridleyi*.

Previously known Distribution. Indian Ocean (?) (auctorum); Aru Islands (Hentschel).

Register Number, Localities, &c. XVII. 1, Kiu, littoral, 24.12.05; XXVI. 5, Adatra Reef, 25 Dec. '05; XXXIII. 7, Dhed Mora and adjacent rocky ground between Beyt and Aramra, 1 fm., 21.12.05.

15. Gelliodes fibrosa Dendy.

Gelliodes petrosioides, var *fibrosa* Dendy [1905].

In my report on Professor Herdman's Ceylon sponges I suggested that my *Gelliodes petrosioides* var. *fibrosa* might, when better material was forthcoming, have to be considered as a distinct species. The occurrence of two fairly good specimens in Mr. Hornell's collection leads me to carry out this suggestion. The two specimens come from the same locality and are possibly parts of the same. One (R.N. XXXIV. 6) is an irregular, flattened sponge which has been attached to the substratum at a few points only, and with a tendency to throw off digitiform processes. The upper surface is almost flat and minutely conulose. It bears several fair-sized but shallow vents, whose margins are very slightly raised above the general surface. The specimen measures about 48 mm. in length by 28 mm. in greatest breadth and 10 mm. in average thickness. Texture rather soft and compressible. Colour in spirit (after formalin) very pale brown. The second specimen is an irregularly subcylindrical fragment (?), about 56 mm. in length by 7 mm. in diameter in the middle, broadening out suddenly to about 18 mm. at one end.

The skeleton arrangement is identical with that of a typical Pachychalina. The main skeleton consists of a subrectangularly meshed network of stout spicular fibre with very little spongin, with numerous scattered oxea in the meshes between the fibres. The dermal skeleton is an irregular network of similar fibre.

Spicules. (1) Slightly curved oxea, gradually and sharply pointed at each end, measuring about 0·19 by 0·008 mm. (2) Slender sigmata, usually simply C-shaped, measuring about 0·0164 mm. from bend to bend.

Previously known Distribution. Ceylon Seas (Dendy).

Register Numbers, Locality, &c. XXXIV. 6, 12, Channel, W. side of S. end of Beyt Island, 3–4 fms., 3.1.06.

16. Reniera permollis (Bowerbank).

Isodictya permollis Bowerbank [1866].

I identify with this species a number of encrusting specimens growing over the parchment-like worm-tubes that are so abundant in the collection. The colour in spirit (after formalin) is rather dark brown ; the texture very soft and friable ; the surface very minutely hispid owing to the projection of the ends of the primary skeletal lines. The oscula are small and scattered. The crusts attain a considerable size, but are so irregular that it is useless to give dimensions.

The skeleton is an irregular isodictyal reticulation, for the most part of single spicules, but there is a strong tendency to form primary lines several spicules thick and separated from one another by intervals of about one spicule's length.

The oxea are gently curved, gradually and sharply pointed, and measure up to about 0·12 by 0·006 mm., agreeing very closely with Bowerbank's figure and measurement.

This seems to be quite a good identification, especially as Bowerbank mentions the " nut-brown colour," but whether *R. permollis* is more than varietally distinct from *R. cinerea* may be regarded as an open question.

Previously known Distribution. British Seas (Bowerbank).

Register Numbers, Localities, &c. XX. 1, Adatra ; XXX. 3, 3¾–4 fms., N. of Poshetra, 20.12.05 ; XXXII. 4, off Beyt.

17. Reniera topsenti Thiele.

Reniera cinerea (Grant) var. *porosa* Topsent [1901].

Reniera topsenti Thiele [1905].

Three well-preserved pieces, possibly parts of the same specimen, agree remarkably closely in external form with Topsent's figure and fairly closely with Thiele's. The sponge is depressed and ramo-lobose in form, with numerous large, slightly prominent vents. The surface has a porous appearance, but is really covered by a thin, almost aspiculous dermal membrane. The texture is rather soft and friable ; the colour in alcohol (after at any rate some formalin) light brown. The skeleton is a rather irregular network of mostly single spicules,

H

but with a tendency to form slender, multispicular primary lines running towards the surface.

The oxea are slightly curved, sharply and fairly gradually pointed, and measure about 0·2 by 0·009 mm., being thus a little larger than those of Topsent's specimens, from which they also differ in the absence of abnormal forms.

Previously known Distribution. Magellan Straits (Topsent) ; Punta Arenas (Thiele).

Register Number, Locality, &c. XXVI. 6, Adatra Reef, 25.12.05. .

18. Reniera hornelli n. sp.—(Plate II., Fig. 11).

The sponge (Fig. 11) consists of an irregularly subglobose body contracted almost (or quite) to a short stalk at the point of attachment. There are numerous large vents, scattered, usually on more or less strongly developed prominences, over the upper parts of the sponge, and varying in diameter up to about 5 mm. Each vent is the terminal opening of a very deep, cylindrical oscular tube. The largest specimen measures about 45 mm. in height by 53 mm. in greatest breadth. The surface of the sponge has a characteristic woolly appearance, due to the fact that the sponge-tissue is broken up into a sort of network of villi by the innumerable narrow, but deep and close-set, inhalant canals. In life the surface was doubtless covered all over by a very thin, translucent, pore-bearing dermal membrane, but this is now nearly all rubbed off. The texture is very soft, spongy and resilient ; the colour in spirit, pale yellowish-grey.

The main skeleton consists of numerous slender, multispicular fibres running at right-angles to the surface at distances of about one spicule's length from one another. These lines are united with one another cross-wise by numerous single spicules and the whole skeleton forms a rather irregular, almost isodictyal network. The dermal membrane appears to be almost aspiculous, and one cannot speak of a definite dermal skeleton. There is very little spongin present in any part of the skeleton.

Spicules. Rather slender, slightly curved oxea, gradually and sharply pointed at each end ; measuring about 0·14 by 0·008 mm., but somewhat variable and often more slender.

I have much pleasure in naming this beautiful and well-characterised species after its discoverer, Mr. James Hornell. As there are four specimens in the collection it is probably not uncommon on the west coast of India.

Register Numbers, Localities, &c. II. 12 (locality uncertain) ; IV. 4, three specimens, dredged off S.W. coast of Beyt Island.

19. Reniera fibroreticulata n. sp.—(Plate II., Fig. 12.)

The sponge (Fig. 12) has the appearance of being made up of short, ana-stomosing branches, sometimes united laterally so as to give a plate-like form, and sometimes, at any rate, ending blindly. Vents relatively large, about 2 mm. in diameter, situated on the sides of the branches or the upper margin of the plate, each the terminal opening of a deep, cylindrical oscular tube. Surface smooth, covered by a closely adherent, translucent dermal membrane, through which the subdermal reticulation shows faintly. Colour in spirit pale yellow, texture fairly firm but friable.

There are several pieces of this sponge in the collection, which, as they all come from the same jar, probably belong to the same specimen. The largest, represented in Fig. 12, is, as a whole, lamellar, measuring about 31 mm. in length, 20 mm. in height and 5 mm. in thickness (this being also about the usual diameter of freely projecting branches).

The main skeleton is a close, irregular network of single spicules, penetrated by long, multispicular fibres about 0·04 mm. thick. These fibres are numerous and run for the most part lengthwise in the branches; they also form a well developed subdermal reticulation, with very irregular, unequal meshes.

The dermal skeleton consists of single spicules, thickly and evenly scattered through the dermal membrane, crossing one another at all angles, but not united in a regular network.

There is little, if any, spongin present.

Spicules. Short and fairly stout oxea, slightly curved and gradually and sharply pointed at each end, measuring about 0·1 by 0·006 mm., but often rather more slender.

This sponge reminds one rather of the European *Reniera simulans,* but differs in the smaller size of the spicules and in the strongly developed reticulation of spicular fibre. In this respect, like *Reniera semifibrosa,* it approaches the genus Pachychalina.

It also resembles, both in external form and spiculation, the new species *Siphonochalina minor,* described below (*cf.* Fig. 15). The growth of the sponge, however, is on a much smaller scale, and the arrangement of the main skeleton is different, for in *S. minor* it is entirely composed of multispicular fibres.

Register Number, Locality, &c. II. 8 (exact locality uncertain).

20. Reniera semifibrosa n. sp.—(Plate II., Fig. 13).

The sponge forms massive, convex crusts, closely adherent to other objects or hollow underneath. The finest specimen (R.N. XXXIII. 1, Fig. 13) has the form of a deep, inverted cup, the margin of which has been broken away all round from the substratum, while the interior is quite hollow and empty. The total height of the specimen is 85 mm. ; the maximum breadth of the base about the same ; the thickness of the wall of the cup about 15 mm. The wall of the cup is perforated by two large, irregular apertures (natural) with rounded margins, which lead right through into the spacious interior.   The inner surface of the cup is smooth but uneven ; it bears no vents. The outer surface is similar, but bears numerous large, prominent, circular vents, up to about 9 mm. in diameter, leading out of wide exhalant canals which come from deep down in the interior of the sponge.   The texture is fairly firm but rather cavernous and friable ; the colour in spirit (after formalin) pale yellowish-grey, with a transparent look which is probably due largely to the imperfect preservation.

Another specimen (R.N. XXXIII. 2 a) forms a much flatter crust growing over a massive specimen of *Jaspis reptans*, and there are also a number of broken fragments evidently of the same species.

The main skeleton is a typical unispicular, isodictyal reticulation of short oxea, and there is a similar unispicular dermal reticulation, except that here the spicules all lie in one plane, parallel to the surface.   Just beneath the surface, however, there is a well developed subdermal reticulation of short, multispicular fibres.   This reticulation lies parallel to the surface.   Its meshes are very irregular in shape and vary greatly in size, and the component fibres vary very much in thickness, up to at least 0·17 mm. The dermal and subdermal skeleton are similar on the inner and outer surfaces of the sponge.   Here and there in the interior of the sponge a reticulation of coarse spicular fibre similar to the subdermal reticulation is to be found, probably representing lines of growth (earlier surface levels).

There is little, if any, spongin present in the skeleton.

Spicules.   Oxea ; fairly stout, slightly curved, sharply and rather abruptly pointed, but not tornote.   Size about 0·16 by 0·0095 mm.

This handsome species, with its well developed subdermal reticulation of stout multispicular fibre, seems to be intermediate between Reniera and Pachychalina.

Register Numbers, Localities, &c. XXXIII. 1, 2 a, 3, Dhed Mora and adjacent rocky ground between Beyt and Aramra, 1 fm., 21.12.05 ; XXXIV. 7, 9, Channel, W. side of S. end of Beyt Island, 3–4 fms., 3.1.06.

Reniera spp.

There are also in the collection a number of more or less fragmentary specimens probably representing other species of this difficult genus.

21. Halichondria panicea Johnston vars.

(For Literature, Synonymy, &c., *vide* Ridley and Dendy [1887] and Dendy [1905].)

There are several specimens in the collection which may be considered as varieties of this ubiquitous species. R.N. XX. 3 a is the best preserved and characterised.   It consists of a compressed lobose fragment (? erect), 51 mm. in length and 30 mm. in greatest breadth (near the top, where it expands somewhat).   A number of good-sized vents occur around the margin.   The surface, under a pocket lens, appears very distinctly reticulate. The texture is firm and rather coarse, the colour in spirit (after formalin) light brown.

The skeleton is a confused reticulation of large oxea which show a strong tendency to arrange themselves in coarse fibres. There is a very well developed dermal reticulation of spicular fibre varying in diameter.

The oxea are slightly curved, gradually and sharply pointed at each end, and commonly measure about 0·77 by 0·023 mm., although variable.

R.N. XXXII. 5 is more massive, with a more compact and less distinctly fibrous skeleton, and comes very near to Ceylon specimens of *Halichondria panicea* var. *megalorhaphis* collected by Professor Herdman [Dendy 1905]. Most of the specimens are in a very poor state of preservation, owing, doubtless, to the use of formalin as a preservative.

Previously known Distribution of the Species. Almost cosmopolitan.

Register Numbers, Localities, &c. XVII. 3, Kiu littoral; XX. 3 a, b, Adatra; XXIX. 4, 3¾–4 fms. N. of Poshetra, 20.12.05 ; XXXII. 5, off Beyt ; XXXIV. 8, Channel, W. side of S. end of Beyt Island, 3–4 fms., 3.1.06.

22. **Halichondria reticulata** Baer [1905].—(Plate II., Figs. 14 a, 14 b.)

The sponge, of which there is a good deal in the collection, consists of slender, irregular, often tortuous, creeping branches, which frequently anastomose and unite with one another in irregular, massive lumps (Figs. 14 a, 14 b). Individual branches are usually about 2 or 3 mm. in diameter. They are generally subcylindrical, but may be flattened ; they may be bluntly rounded at the extremity or drawn out into a long point. The surface is finely granular and marked by ramified, meandering, subdermal canals. Vents minute, few, scattered. The colour in spirit is very pale yellow ; the texture rather compact and fleshy, but soft and compressible.

The main skeleton consists of slender oxea, partly scattered quite irregularly and partly in loose wisps or fibres which run towards the surface. The dermal skeleton consists of irregularly scattered spicules of the same kind, lying tangentially and crossing one another in all directions.

Spicules. Slender oxea ; very slightly curved, gradually sharp-pointed at each end, measuring about 0·21 by 0·006 mm. ; very uniform in shape and size.

This species seems to be well characterised by its peculiar mode of growth and by the small size of its spicules. Although Baer's description is very brief, and although the spicules in the Okhamandal sponge seem to be somewhat stouter, I think the identification is a fairly safe one. Baer observes that the flagellate chambers are round and 0·03 mm. in diameter, and that the ground substance is homogeneous and filled with round granule-cells.

The histological features of the Okhamandal sponge, which I have studied by means of paraffin sections of material stained with borax-carmine, confirm the identification with Baer's species. The flagellate chambers are approximately spherical and about 0·03 mm. (or a little more) in diameter, scattered rather sparsely in a compact ground-substance densely charged with minute, spherical, granule-bearing cells, measuring up to about 0·008 mm. in diameter. There is a fairly thick ectosome, more or less interrupted by the spacious subdermal cavities and also containing small granule-cells.

Throughout the sponge, but especially in parts of the ectosome, where they are densely crowded together, occur numerous large, spherical cells, filled with granules of various sizes. and each with a compact, deeply staining nucleus of moderate size. The diameter of the entire cell is about 0·025 mm. ; of the nucleus about 0·008 mm. These cells resemble immature ova, but they may be merely large amœbocytes charged with food-material or excretory products.

Although Baer has figured both external form and spicules in the case of the Zanzibar sponge, I have thought it desirable to add illustrations of the Indian form.

Previously known Distribution. Zanzibar (Baer).

Register Numbers, Locality, &c. II. 6, 11 (altogether a considerable number of pieces), probably off Poshetra, January 7, 1906.

23. Siphonochalina crassifibra Dendy.

Siphonochalina crassifibra Dendy [1889].
Siphonochalina communis (pars) Dendy [1905].

The best of the two specimens in the collection closely resembles in external form Carter's *Siphonochalina (Patuloscula) procumbens* from the West Indies, a figure of which will be found in my memoir on the West Indian Chalininæ [1890]. It consists of a spreading base from which a dozen or more tubes rise obliquely upwards, branching and anastomosing with one another to a slight extent. The tubes are subcylindrical, about 50 mm. in height and 12 or 14 mm. in diameter, and each terminates in a wide, circular vent about 6 mm. in diameter. They are thus considerably smaller than in the type. The surface is smooth but finely granulated.

In skeletal peculiarities the specimens exaggerate the distinguishing character of the type. The main skeleton is a rectangularly or polygonally meshed network of fairly stout fibre, almost completely filled with the very numerous, close-packed spicules, so that there is only a thin investment of spongin. The primary fibres are about 0·05 mm. in diameter and the secondaries only a little less. The dermal skeleton in the best specimen, from which this description is taken, is chiefly a unispicular reticulation, in which the spicules are held together by very pale-coloured spongin, with a much coarser subdermal reticulation formed by the outer part of the main skeleton. The spicules are very slightly curved, fairly sharply pointed oxea, measuring about 0·078 by 0·004 mm.

In my Report on Professor Herdman's Ceylon Sponges [1905] I suggested that my *Siphonochalina crassifibra* should be regarded as a variety of Carter's *Siphonochalina* (*Tubulodigitus*) *communis*, also an Indian Ocean species. Probably we shall ultimately have to unite in one species a considerable number of varieties which exhibit the same characteristic external form, *e.g.*, *Tubulodigitus communis* Carter [1881], *Siphonochalina crassifibra* Dendy [1889], *Siphonochalina communis* var. *tenuispiculata* Dendy [1905], from the coasts of India and Ceylon; *Patuloscula procumbens* Carter [1882, 1885 *bis*], from Australia and the West Indies; *Siphonochalina intermedia* Ridley and Dendy [1887], from Australia; *Siphonochalina spiculosa* Dendy and *S. ceratosa* Dendy [1890], from the West Indies. In all these localities there seems to be much variation in the relative amounts of spicules and spongin in the skeleton fibres. There is also a good deal of variation in the length and diameter of the tubes, but I do not see how it is feasible to make really satisfactory specific distinctions between the forms mentioned.

Until, however, it is possible to make a more thorough comparative study of this interesting group of varieties I propose to revert to the specific name " *crassifibra* " for the Indian form in question.

Previously known Distribution. Gulf of Mannar (Dendy).

Register Numbers, Localities, &c. VII., Adatra Reef, 5.12.05; XXXIV. 5, Channel W. side of S. end of Beyt Island, 3–4 fms., 3.1.06.

24. Siphonochalina minor n. sp.—(Plate II., Fig. 15.)

The sponge (Fig. 15) consists of a horizontal, subcylindrical or vertically somewhat flattened, branching and anastomosing stolon, from which arise short, slightly branching and sometimes anastomosing, ascending tubes, terminating each in a wide vent. In the best specimen (R.N. XXVI. 4) there is a continuous stolon 53 mm. in length by about 6 mm. in diameter; the largest tube arising from it is about 35 mm. in height and 10 mm. in average diameter, with a terminal vent about 3·5 mm. in diameter. The stolon itself, though penetrated by smaller canals, does not contain a single, wide, central cavity, as do the tubes that arise from it. The surface is subglabrous, slightly uneven, very minutely granular. The texture is very compressible and resilient, easily torn. The colour in spirit (after formalin) is light brown.

The main skeleton is a rectangularly meshed network of stout multispicular fibre, the meshes becoming irregularly polygonal in the deeper parts. Primary and secondary fibres are alike and of about the same diameter, 0·034 mm. They are composed of a great number of oxea and a small quantity of spongin, which does not seem to form a continuous investment but is sometimes visible in the angles of the reticulation. The dermal skeleton is an irregular, unispicular reticulation of oxea, in which the individual spicules overlap one another extensively ; without obvious spongin.

The spicules are very slightly curved oxea, gradually and sharply pointed at each end, measuring about 0·13 by 0·006 mm.

This species shows very clearly the difficulty of differentiating between the genera Pachychalina and Siphonochalina. The whole structure of the stolon is typically that of a Pachychalina, and only the presence of the tubular ascending branches justifies its inclusion in Siphonochalina. The entire sponge resembles *Siphonochalina crassifibra* on a smaller scale, but with decidedly larger spicules. It makes a near approach to my *Pachychalina subcylindrica* from Ceylon [1905]. I have already pointed out the resemblance which it bears to *Reniera fibroreticulata* (*vide supra* and *cf.* Fig. 12).

Register Numbers, Localities, &c. XXVI. 4, Adatra Reef, 25 December, '05 ; XXXIII. 5, Dhed Mora and adjacent rocky ground between Beyt and Aramra, 1 fm., 21.12.05.

Chalininae spp.

There are several other small chalinine sponges in the collection which are not in a sufficiently good state of preservation or sufficiently well characterised to make identification or description desirable.

25. Desmacella tubulata Dendy [1905].

It is very interesting to meet with this curious and well characterised species again in Mr. Hornell's collection. So far as I am aware it has not been recorded since I first described it from Ceylon in 1905. The material now before me consists of a number of fragments of thin-walled tubes, some of which seem to have been 10 mm. or more in diameter. At one point some of these tubes have grown into close union with a specimen of Gellius sp.

The agreement in spiculation with the type is very close, but I must add that I have observed a few slender toxa, about 0·032 mm. long, on one occasion arranged in a sheaf or toxodragma, differing only in the curvature of the toxa from the smaller trichodragmata. I did not observe any toxa in the type specimens, but probably they occur there amongst the vast number of trichodragmata. I must also add to my original description that the individual trichites of which the trichodragmata are composed, although slender, are not nearly so slender as in some cases. The sheaves or dragmata themselves may be very thick.

Previously known Distribution. Gulf of Mannar (Dendy).

Register Number, Locality, &c. XXVI. 7, Adatra Reef, 25 December, '05·

26. **Thrinacophora cervicornis** Ridley and Dendy.

Thrinacophora cervicornis Ridley and Dendy [1887].

Thrinacophora cervicornis Hentschel [1912].

This species is represented in the collection by two small specimens. The branching, so far as the specimens show, is dichotomous and in one plane, the branches being very short and bluntly rounded off. Surface sparsely hispid owing to the long, slender styli, which project for nearly two millimetres, with shorter projecting styli between. The two specimens are very similar to one another in appearance and closely resemble the young antlers of a stag " in velvet." Total height of each about 20 mm.; diameter of main stem and branches about 3·5 mm., but variable. Colour in spirit light brown.

There is a very stout skeletal axis composed of a dense reticulation of stout and rather short oxea. This is surrounded by a comparatively thin layer of soft tissue in which loose bundles of very long styli run lengthwise. Similar styli, with their bases implanted in the central axis, extend through the outer layer and for a long distance beyond the surface. These are arranged singly and each is surrounded, where it leaves the surface, by a radiate tuft of comparatively short and very slender styli. There is no conspicuous spongin in any part of the skeleton (in balsam preparations).

Spicules. 1. Oxea of the central axis; rather short, stout, distinctly but not very strongly curved or angulated in the middle; usually fairly gradually and sharply pointed at each end. Size fairly uniform, about 0·26 by 0·013 mm.

2. Large styli, running lengthwise in the outer part of the sponge and also projecting more or less at right angles from the surface. These spicules are stoutest at the base and from there taper very gradually to very fine points. They measure about 2·24 mm. in length by 0·02 mm. in thickness at the base. They are slightly curved.

3. Small styli of the radiate surface-tufts; very slender and finely pointed; measuring about 0·55 by 0·0043 mm.

4. Trichodragmata; occurring in immense numbers in the outer part of the sponge; each bundle rather long and narrow, measuring about 0·1 by 0·0082 mm., often curved, easily separating out into trichites.

In spite of some slight apparent differences in spicular proportions, especially as regards the smaller size of the large styli, I think that there can be no reasonable doubt that we have here two small and probably young examples of the *Challenger* species described by Mr. Ridley and myself under the name *Thrinacophora cervicornis*.

In the *Challenger* Report the length of the trichodragmata is given as 0·0126 mm. It should be 0·1 mm., as determined by re-measurements of the type.

I believe that this remarkable and well-characterised species has only been recorded once since it was first described in the *Challenger* Report. *Previously known Distribution.* Philippine Islands (Ridley and Dendy); Aru Islands (Hentschel). *Register Number, Locality, &c.* III. 6 *a*, *b*, dredged off Dwarka.

27. Axinella virgultosa Carter [1887].

There is in the collection one remarkably beautiful specimen, which agrees so closely in its very characteristic external form with the description and figure given by Carter of his *Axinella virgultosa* from Mergui that I have little doubt in making a specific identification. Carter's description of the spiculation, however, without either figures or measurements, is so inadequate that it is impossible, without referring to the type, presumably in Calcutta, to be quite certain. Unfortunately, there is no microscopic preparation of the sponge in Mr. Carter's cabinet.

Mr. Hornell's specimen consists of a large number of slender, vertical, stiff processes, rising up side by side from a thin, encrusting base which measures about 35 by 22 mm. The larger processes are in the middle and measure about 20 mm. in height. They are about 2 mm. in diameter at the base and taper gradually to sharply pointed apices. The processes may give off branches, chiefly from their outer sides near the base. They are all abundantly but shortly hispid with projecting spicules. The colour in spirit (after formalin) is almost white.

Little more than the skeleton of the sponge remains. Each process consists of a plumose column of short, stout spicules arranged in typically axinellid fashion.

The spicules are typically stout styli, sometimes becoming oxeote by more or less pronounced sharpening of the inner end. They are commonly a little bent and the outer end is gradually sharp-pointed. I have seen nothing of the subterminal inflation which Mr. Carter says is often present in his specimens. These spicules measure about 0·77 by 0·04 mm.

Previously known Distribution. Mergui Archipelago (Carter).
Register Number, Locality, &c. IX., off Dwarka, 15–17 fms., 12.12.05.

28. Phakellia donnani (Bowerbank).
Isodictya donnani Bowerbank [1873].
Axinella donnani Dendy [1887].
Phakellia donnani Dendy [1905].
Phakellia donnani Row [1911].

This well-known Ceylon species is represented in the collection by a single specimen of the ordinary, pedunculate, cup-like form. The colour in life was, as usual, orange. It differs from my Ceylon specimens only in the somewhat larger, and especially stouter, spicules, which measure, when full-grown, about 0·38 by 0·025 mm.

Previously known Distribution. ˙ Gulf of Mannar, Ceylon Seas (Bowerbank, Dendy); Red Sea (Row).

Register .Number, Locality, &c. II. 7, Poshetra Head, 7.1.06.

29. Auletta lyrata var. glomerata Dendy.—(Plate II., Fig. 16).

Spongia lyrata Esper [1794–1806].
Raspaigella lyrata Ehlers [1870].
Auletta aurantiaca Dendy [1889].
Auletta lyrata var. *glomerata* Dendy [1905].

The single specimen (Fig. 16) in the collection agrees very closely in general form with the type of the variety. The strongly sphinctrate vents are situated each in a cup-shaped depression at the extremity of a short branch. The average thickness of the spicules is, however, much greater than in the type of the variety.

The species of the genus Auletta are evidently extremely variable both in external form and spiculation, and it will be extremely difficult to differentiate them from one another.

Previously known Distribution of Species. Ceylon, Gulf of Mannar (Esper, Dendy).

Register Number, Locality, &c. III. 3, off Dwarka.

30. Auletta elongata Dendy var. fruticosa nov.—(Plate II., Fig. 17).

Auletta elongata Dendy [1905].

The single specimen (Fig. 17) differs from the type in its much more spreading mode of branching and in the considerably smaller average. size of the spicules.

Previously known Distribution of Species. Ceylon, Gulf of Mannar (Dendy).

Register Number, Locality, &c. XXIII. 2, off Dwarka, 15–17 fms., 12.12.05.

31. Ciocalypta dichotoma n. sp.—(Plate III., Fig. 18).

The single specimen (Fig. 18) consists of a cylindrical stem dividing at half the total height of the specimen into two approximately equal branches diverging from one another at an acute angle. Each branch terminates in a bluntly pointed apex. The base of the stem is slightly enlarged and attached to it are a few grains of coarse sand and a comparatively large shell-fragment, which seem to indicate that the stem was directly attached to the substratum and did not spring from a massive body. The surface is stellately reticulate as in *Ciocalypta hyaloderma* Ridley and Dendy [1887], though hardly so distinctly. There is a translucent dermal membrane, supported by a reticulation of spicular fibre and overlying extensive subdermal cavities. There are a number of small, inconspicuous vents, chiefly in single series along the sides of stem and branches. Colour in spirit very pale yellow; texture stiff, resilient. Total height of specimen 46 mm.; diameter of stem and branches about 4–5 mm.

The sponge consists, as usual in the genus, of a central axis, surrounded by wide subdermal cavities which are traversed by spicular columns supporting the dermal membrane. The axis is very thick and the radiating spicule columns very short. Numerous loose fascicles of large oxea run lengthwise through the axis, separated from one another by a fair amount of soft tissue and crossed here and there by scattered oxea. The radiating columns which support the dermal membrane are loose fascicles of similar spicules, ending in surface brushes of short styli. The dermal skeleton is a very irregular reticulation of loose spicular fibre composed of the large oxea. In the dermal membrane also occur numerous small styli, mostly arranged in the above-mentioned brushes at the ends of the radial columns.

The spicules are very sharply differentiated into two kinds :—(1) Large oxea ; slightly curved, fusiform, symmetrical, gradually and sharply pointed at each end, size about 0·8 by 0·02 mm. (smaller ones also occur). (2) Small styli ; short, usually slightly bent ; well rounded off, but somewhat narrowed, at one end, and gradually sharp-pointed at the other, size about 0·2 by 0·008 mm.

The spiculation of this sponge seems to be identical with that of my *Hymeniacidon* (?) *fœtida* [1889], originally from the Gulf of Manaar, which has been included by Lindgren [1898], Thiele [1900] and Hentschel [1912] in the genus *Ciocalypta*, probably quite rightly. The external form of *Ciocalypta dichotoma*, however, is so definite, and so different from that of *C. fœtida*, that I think they may, for the present, be regarded as distinct, though closely related species. It should be borne in mind, on the other hand, that Hentschel [1912] describes specimens of *C. fœtida* with finger-shaped processes.

Register Number, Locality, &c. IV. 21, dredged off S.W. Coast of Beyt Island.

32. Higginsia sp.

I have no hesitation in referring to this genus a subcylindrical fragment measuring about 30 by 8 mm. Unfortunately, the specimen was preserved in formalin, and is in a very badly macerated condition, practically nothing but the skeleton remaining, while it contains a large number of evidently foreign spicules.

The main skeleton is a very irregular, subfibrous reticulation of large, stout oxea, measuring about 0·9 by 0·03 mm., only very slightly curved and gradually and sharply pointed. These are accompanied by a number of very much longer and very much more slender oxea and styli, which seem to belong to the sponge.

There are two kinds of microxea; (1) covered with small sharp spines; more or less sharply angulated in the middle; size about 0·15 by 0·006 mm.; (2) smooth, but with a distinct swelling at one side of the central angulation (subcentrotylote); of about the same length as the spined ones but much more slender, even when allowing for the absence of spines. An intermediate form may occasionally be found, but on the whole the two kinds seem to be fairly distinct. It is possible, however, that the smooth ones may be merely young forms of the other.

I refrain from giving a name to this species until better material is forthcoming.

Register Number, Locality, &c. XVIII. 4 *a*, Channel W. of Beyt Island, 3–4 fms., January 7, '06·

33. **Esperella plumosa** (Carter).—(Plate I., Figs. 4*a*–4*g* ; Plate III., Fig. 19).
Esperia plumosa Carter [1882, 1887].
Esperella plumosa Dendy [1905].
Not *Esperella plumosa* Arnesen [1903].

This appears to be a very common and characteristic Indian Ocean species and is by far the most abundant sponge in Mr. Hornell's collection. It was also abundant in Professor Herdman's Ceylon collection. As yet, however, no figures, either of the external form or of the very well developed spiculation, have been published, and it seems desirable to make good this omission on the present occasion.

The external form appears to be very characteristic, though varying much according to the stage of growth. The sponge seems to begin life as an irregular crust, which becomes ·massive and then grows out into long, flattened, tongue-shaped processes, finally breaking up into slender, pointed, digitiform branches; or possibly such branches may be formed first as outgrowths of the massive crust, and subsequently fuse to form the flattened tongue-shaped portions. One of the best pieces is represented of the natural size in Fig. 19, but this is not the largest specimen.

The irregularly conulose or cactiform surface also seems to be characteristic. The surface is subglabrous between the conuli and the dermal reticulation of spicular fibre appears to be very unequally developed; in one specimen which I have dried it is quite conspicuous in some places, under a pocket lens, while apparently absent in others. The oscula are represented by larger and smaller circular apertures scattered here and there between the conuli.

Most of the specimens are now, in spirit and dry, of a dull reddish colour, but I suspect that this may be due to their having been preserved in the first instance in formalin.

I have nothing further to add to my previous description of the skeleton arrangement, but a more complete account of the spiculation seems desirable to accompany the illustrations, especially as I omitted before to mention two types of spicule which occur both in Carter's type and in the Ceylon and Okhamandal specimens. These are the small, palmate anisochelæ and the small, slender sigmata. I think I must have regarded these formerly as merely young forms, but I do not think that that view can be accepted. There are, then, no fewer than seven different kinds of spicule in the sponge, and the constancy in form and size in all the specimens I have examined, including the type, is very remarkable.

1. Tylostyli (Fig. 4a), of the usual Esperella type, generally slightly crooked ; size about 0·3 by 0·009 mm. (much more slender forms also occur).

2. Large, broad, palmate anisochelæ (Figs. 4b–4b″), about 0·049 mm. in length by 0·022 mm. in greatest width from one lateral palm to the other.

3. Small palmate anisochelæ (Figs. 4c, 4c′), about 0·02 mm. long by 0·008 mm. in greatest width.

4. Minute palmate isochelæ (Figs. 4d, 4d′), resembling those of Clathria, about 0·012 mm. long.

5. Large, stout sigmata (Figs. 4e, 4e′), probably all really more or less contort, with abruptly recurved and very sharply pointed ends ; length in a straight line from bend to bend about 0·094 mm., thickness about 0·0054 mm.

6. Small, slender, contort sigmata (Fig. 4f) ; length in a straight line from bend to bend about 0·033 mm., thickness about 0·0013 mm. ; but variable in dimensions.

7. Slender toxa (Fig. 4g), gently curved like a parenthesis mark ; length about 0·065 mm., thickness about 0·0013 mm. ; sometimes arranged in toxodragmata.

Pending a much-needed revision of the esperelline sponges I adhere to the genus Esperella for this species, but, apart altogether from the question whether or not that name should be replaced by Mycale, it seems probable that the genus will have to be split up in the near future.

I have already [1905] pointed out the close resemblance of the characteristic large anisochelæ of this species to the corresponding spicules of *Esperella simonis* Ridley and Dendy [1887], from Simon's Bay, Cape of Good Hope. It is obvious that the two species are closely related, but there are certain well-marked differences in the spiculation. Chief amongst these are the absence from the spiculation of *E. simonis* of the minute, palmate isochelæ and of the small, slender sigmata, both of which are very abundant in *E. plumosa*, which has an extraordinarily full complement of spicules. Other differences concern the size of the spicules and the shape of the toxa, which are much more strongly arcuate in *E. simonis*.

Previously known Distribution. Mauritius and Mergui Archipelago (Carter) ; Ceylon (Dendy).

Register Numbers, Localities, &c. XX. 2, 8, Adatra ; XXII. 1, 3, ½ mile N. of Poshetra, 20.12.05 ; XXIV. 1, 2, Kiu, low water, 24.12.05 ; XXVI. 3, 12, Adatra Reef, 25 Dec. '05 ;ᠯ XXXII. 1, 2, 7, off Beyt ; XXXIV. 3 *a*, channel, W. side of S. end of Beyt Island, 3–4 fms., 3.1.06.

34. Desmacidon minor n. sp.—(Plate III., Figs. 20*a*, 20*b*.)

There are two specimens of this sponge in the collection, both from the same locality. One of them (R.N. XX. 6 *a*, Fig. 20*a*) is compressed, flabellate, with a much contracted base, almost forming a short peduncle. The upper margin of the sponge is widely extended and produced into short conical processes, apparently the ends of laterally fused branches. The vents are of fair size, shallow and scattered on one surface of the sponge only, not marginal ; owing to the maceration of the specimen they are no longer very distinct. The surface is finely granular. The specimen measures about 26 mm. in greatest height, 41 mm. in width, and 3 mm. in average thickness. The texture is soft and resilient ; the colour in spirit (after formalin), light brown.

The second specimen (R.N. XX. 6 *b*, Fig. 20*b*) is very similar in most respects, but is divided into digitiform branches from the contracted base upwards, most of the branches lying in approximately the same plane. The branches are subcylindrical or slightly compressed, and about 3–6 mm. in diameter. The shallow vents tend to arrange themselves in longitudinal series on the branches. The maximum height of the specimen is about 33 mm.

The skeleton is a sub-isodictyal reticulation of short oxea, but with the meshes composed of plurispicular fibres in which the spicules are held together by a considerable amount of very pale-coloured spongin.

The spiculation consists of the following :—

(1) Fairly stout, slightly curved oxea, gradually and sharply pointed at each end, measuring about 0·13 by 0·008 mm. Much more slender oxea also occur, which are probably young forms.

(2) Slender, palmate isochelæ (" naviculiform "), about 0·0164 mm. long ; abundant.

This species might almost be regarded as a dwarf variety of *Desmacidon compressa* (Carter's *Chalina compressa*), which appears to be a South African species, but the spicules of the latter are very much larger, the oxea measuring, according to Carter [1882 *bis*], about 0·37 by 0·023 mm, and the isochelæ (" naviculiform equianchorates ") 0·025 mm. in length.

Register Numbers, Locality, &c. XX. 6 *a, b*, Adatra.

35. Iotrochota baculifera Ridley.

Iotrochota baculifera Ridley [1884].
Iotrochota baculifera var. *flabellata* Dendy [1887].
Iotrochota baculifera Topsent [1893].

Iotrochota baculifera Topsent [1897].
Iotrochota baculifera Lindgren [1898].
Iotrochota baculifera Thiele [1899].`
Iotrochota baculifera var. *tumescens* Kirkpatrick [1900].
Iotrochota baculifera Thiele [1903].
Iotrochota baculifera Dendy [1905].
Iotrochota baculifera var. *minor* Hentschel [1911].
Iotrochota baculifera Hentschel [1912].

There is in the collection only a single specimen of this widely distributed and well known Indian Ocean species. It forms a rather thick, irregular crust of a dark brownish-purple colour, attached to a calcareous nodule, the greatest diameter of the specimen being about 30 mm. The specimen was preserved in alcohol and is in good condition. The skeleton arrangement and spiculation are typical. The styli measure about 0·13 mm. in length by from 0·004 to 0·008 mm. in thickness. The diactinal megascleres are strongylote or have only very feebly developed heads; they measure about 0·2 by 0·004 mm. The "birotulates" are very abundant but very minute, only about 0·0125 mm. long. The spicules therefore are decidedly smaller than in Professor Herdman's Ceylon specimens.

Previously known Distribution. North Australia and Mascarene Islands (Ridley); Gulf of Mannar and Ceylon Seas (Dendy); Seychelles and Amboina (Topsent); Coast of Cochin China (Lindgren); Celebes and Ternate (Thiele); Christmas Island (Kirkpatrick); S.W. Australia and Aru Islands (Hentschel).

Register Number, Locality, &c. XIII., Adatra Reefs, 25 December, '05·

36. **Guitarra indica** n. sp.—(Plate I., Figs. 5*a*–5*b*⁗; Plate III., Fig. 21).

This very interesting sponge is represented in the collection by eight good specimens. Five of these (Fig. 21) are attached to a branching, parchment-like tube, belonging to some polychæte worm*, along with other sponges, including an ·Esperella and an encrusting Aplysillid. Two other loose specimens, in the same jar, have probably been broken away from the same association. The eighth specimen (R.N. II. 4) is also loose and comes from a different locality.

The specimens are irregularly cushion-shaped and tend to surround by overgrowth the object to which they are attached; they bear a general resemblance, in form and colour, to some species of Chondrilla.

The largest measures about 28 mm. in maximum diameter, with a true thickness (from the outer surface to the surface of attachment) of about 7 mm. The colour, in spirit, ranges from slate grey (on the surface which was evidently exposed to the light) to pale yellow (on what was evidently the shaded surface). R.N. II. 4 is pale yellow all over.

* Probably *Eunice tubifex*, see p. 96.

The vents are numerous and irregularly scattered on the more exposed parts of the surface. They are mostly minute and each at the summit of a small conical projection, formed by a contracted spicular, membranous margin, but I have seen one expanded up to nearly 2 mm. in diameter. The surface is smooth and appears porous under a pocket lens, but I have not been able to detect the actual inhalant pores. These, however, are no doubt scattered in the thin dermal membrane between the surface-brushes of megascleres.

The main skeleton is a very irregular, rather close-meshed reticulation of somewhat loose spicular fibre, with a certain number of isolated megascleres scattered between. At the surface this gives place to a velvety pile composed of well defined brushes of megascleres with outwardly directed apices projecting for a short distance beyond the surface. I have detected no spongin.

Spicules. (1) Styli (tornostrongyla) (Fig. 5*a*); nearly straight, rather abruptly sharp-pointed at one end and rounded off at the other; often a little crooked; measuring about 0·266 by 0·007 mm., but often more slender.

(2) Placochelæ (Figs. 5*b*–5*b''''*); of the typical Guitarra form but with the shaft very abruptly constricted in the middle; length about 0·041 mm., with greatest breadth of expanded shaft about 0·0143 mm. Numerous smaller forms occur with less sharp constriction in the middle of the shaft; also numerous very slender forms of various sizes, without fimbriæ or with very feebly developed fimbriæ, which I take to be early developmental stages. The placochelæ are abundantly scattered throughout the choanosome.

So far as I am aware, only three species of Guitarra have hitherto been described; viz. *Guitarra fimbriata* Carter [1874], from deep water in the North Atlantic, *Guitarra voluta* Topsent [1904], from deep water off the Azores, and *Guitarra antarctica* Hentschel [1914 ?], from deep water in the Antarctic. Unfortunately, no depth was recorded for our new species, but it was associated with typical shallow water sponges and the depth was probably not more than a few fathoms.

Guitarra indica differs from all its congeners in the comparatively small size of the placochelæ. It seems to come nearest to *G. voluta*, but differs from that species in the smaller size of the spicules, the presence of the surface pile of megasclere-brushes (in which it agrees with *G. fimbriata*), and in the abrupt constriction of the shaft of the placochelæ. It differs from *G. fimbriata* and agrees with *G. voluta* in having torno-strongylote megascleres (styli), and it differs from *G. antarctica* in having no sigmata. The youngest stages (Fig. 5*b'''*) in the development of the placochelæ, however, resemble sigmata, but they have a peculiar, indefinite, rough outline which seems to indicate immaturity.

Locality, Register Numbers, &c. II. 4, off Poshetra, 7 January '06; IV. 5 *a–g*, dredged off S.W. coast of Beyt Island.

Genus **Psammochela** n. g.

Desmacidonidæ with a reticulate skeleton composed of sandy and sometimes partly spicular fibres. Megascleres styli or strongyla or both. Microscleres isochelæ, which may be very minute and with vestigial teeth ; to which sigmata may be added.

I propose this genus for some interesting sand-sponges from the neighbourhood of Beyt Island, which seem to represent a stage in the regressive evolution of such genera as Phoriospongïa and Chondropsis. These two genera have no chelæ, and I have hitherto excluded them from the Desmacidonidæ, but the occurrence of the vestigial isochelæ in Psammochela seems to indicate à probable desmacidonid origin.

I regard *Psammochela elegans* n. sp. as the type of the genus. Hentschel's *Desmacidon psammodes* from S.W. Australia [1911] evidently belongs to the same genus. My own *Desmacidon* (?) *arenifibrosa* [1896] and Carter's *Dysidea chaliniformis* (= *Desmacidon* (?) *chaliniformis* Dendy [1896]), may belong to a closely related genus in which the megascleres have been completely suppressed.

37. **Psammochela elegans** n. sp. (Plate I., Figs. 6a–6e ; Plate III., Figs. 22a, 22b).

Sponge (Figs. 22a, 22b) irregular, lamellar or digitate, often running out into long, slender, sometimes bifurcating processes, from 3 to 10 mm. in diameter and up to about 90 mm. in length in the specimens before me. Surface irregularly rugose or conulose, but with the intervals between the rugæ or conuli spanned over in life by a delicate, translucent, minutely reticulate and finely porous dermal membrane. In formalin specimens (Fig. 22b) the dermal membrane has completely disappeared, and the sponge has a curiously eroded appearance, the surface being deeply and irregularly grooved and pitted,the grooves and pits being obviously uncovered subdermal cavities. Vents probably rather small and ṣcattered.

The main skeleton is a fairly close-meshed but very irregular reticulation of rather slender fibre, composed of sand-grains and proper megascleres in varying proportions, with no visible spongin ; with numerous megascleres and sand-grains scattered in the soft tissues between the fibres. The dermal skeleton is composed chiefly of very fine, scattered sand-grains, with a more or less pronounced tendency to arrange themselves in a fine-meshed reticulation.

Spicules. (1) Slender styli (Fig. 6a) ; evenly rounded off at the base, rather abruptly sharp-pointed at the apex, often a little crooked ; size about 0·16 by 0·005 mm., but very variable in thickness (? sometimes becoming strongylote). (2) Tridentate isochelæ (Figs. 6b, 6b') ; fairly robust, with stout, curved shaft ; length about 0·024 mm. These are usually scarce. (3) Very minute, C-shaped isochelæ (Figs. 6c, 6c') ; resembling in side view slender, strongly curved, C-shaped sigmata with slightly enlarged ends ; in front view like an Iotrochota " birotulate," with indications of three vestigial teeth at each end ; length from bend to bend about 0·012 mm.

Very numerous. (4) Slender or fairly stout, more or less contort sigmata (Figs. 6d, 6d'); with short, strongly recurved, finely ·pointed ends; measuring commonly about 0·033 mm. from bend to bend. Much smaller sigmata also occur (Fig. 6e), which may belong to a different category.

Register Numbers, Localities, &c. IV. 12, 16, 20, dredged off S.W. coast of Beyt Island; X., XXXIV. 11, channel, W. side of S. end of Beyt Island, 3–4 fathoms, 3.1.06; XVIII. 3, channel, W. of Beyt Island, 3–4 fathoms, Jan. 1906; XXXIII. 6, Dhed Mora and adjacent rocky ground between Beyt and Aramra, 1 fathom, 21.12.05.

38. Chondropsis kirkii (Carter).

Dysidea kirkii Carter [1885].
? *Sigmatella australis* Lendenfeld [1889].
Sigmatella corticata Lendenfeld [1889].
Chondropsis kirkii Dendy [1895].

I identify with this species a single well preserved specimen which differs in no important respect from the common Australian form. The specimen is subdigitate, consisting of two or three tubular processes partially fused together side by side and each terminating in a conspicuous but constricted vent, surrounded by a membranous collar. The surface is slightly conulose, minutely reticulate where rubbed, sub-glabrous where uninjured. The texture is rather soft and compressible, the colour in spirit pale grey. Height of specimen 31 mm., breadth 36 mm., diameter of digitiform processes about 12 mm.

The main skeleton consists of the usual irregular reticulation of fine-grained sand-fibre. The dermal sand-reticulation is less strongly developed than is usual in Australian specimens. The spicules are the usual slender strongyla and small sigmata, the strongyla perhaps rather better developed than in most Australian specimens. ·

Previously known Distribution. Australian Seas (Carter, Lendenfeld, Dendy).

Register Number, Locality, &c. IV. 11, dredged off S.W. coast of Beyt Island.

39. Myxilla arenaria Dendy [1905].

I identify with this species three specimens in the collection, all from about the same locality. The one (R.N. IV. 2 a) upon which the following notes are based is irregularly and massively lobose, slightly clathrous, about 66 mm. in height and 50 mm. in greatest breadth. It looks (in spirit) like a mass of sand held together by pale grey, gelatinous " sarcode," which forms a thicker or thinner surface layer. The surface is uneven and slightly conulose or rugose. The vents are rather small and arranged in a row on the prominent ridges which form the top of the sponge.

The skeleton is composed chiefly of a dense agglomeration of coarse sand-grains which, to the naked eye, show, through the translucent dermal membrane, a distinct tendency to be arranged in ascending columns. These sand-grains are very sparsely echinated by small spined styli, while slender diactinal megascleres occur scattered and in loose wisps in the soft tissues between. Towards the surface the diactinal megascleres become much more abundant and form radiating dermal brushes. The megascleres are rather more robust than in the type and the diactinal megascleres are distinctly tornote, *i.e.* very abruptly pointed at each end, rather than rounded off.

Previously known Distribution. Gulf of Mannar, Ceylon Seas (Dendy).

Register Numbers, Locality, &c. IV. 2 *a, b*, dredged off S.W. coast of Beyt Island ; XXXIV. 4, channel, W. side of S. end of Beyt Island, 3–4 fathoms, 3.1.06.

40. Clathria corallitincta Dendy.

Clathria corallitincta Dendy [1889].
Clathria frondifera Dendy [1905].
? *Clathria frondifera* Ridley [1884].
? *Halichondria frondifera* Bowerbank [1875].

There are three good-sized specimens and one small specimen of this sponge in the collection. They have, each as a whole, a massive form with a slight tendency to become lamellar, but each is in reality made up of an immense number of slender, anastomosing branches, forming a close network, with the ends of the branches projecting on the surface in the form of conuli. All three specimens exhibit large pseudoscula on prominent parts of the sponge.

The spiculation calls for no special comment except as regards the presence of numerous large, stout, strongly bent toxa.

It is quite possible that Bowerbank's *Halichondria frondifera* may be a very variable and widely distributed species, with which the Indian specimens might be identified, as I did with Ceylon specimens in 1905 ; but 1 am now more inclined, in view of Bowerbank's original description and the new material to hand, to regard the Indian and Ceylon specimens as belonging to a distinct species, as I did in 1889, under the name *Clathria corallitincta.*

Previously known Distribution. Gulf of Mannar, Ceylon (Dendy).

Register Numbers, Localities, &c. IV. 14, off S.W. coast of Beyt Island ; XVIII. 2, channel, W. of Beyt Island, 3–4 fathoms, January, 1906 ; XIX. 1, Vamiani Point, January 5, 1906 ; XXXIV. 2, channel, W. side of S. end of Beyt Island, 3–4 fms., 3.1.06.

41. Clathria spiculosa Dendy.

Rhaphidophlus spiculosus Dendy [1889].
Clathria spiculosa Dendy [1905].
Clathria spiculosa var. *ramosa* Dendy [1905].
Clathria spiculosa vars. *ramosa* and *macilenta* [Hentschel 1912].

There are several specimens in the collection which I identify with this common

Ceylon species. The branching is extremely irregular, and the branches vary in form from short, broad and flattened to slender, long and cylindrical. The long, slender-branched specimens evidently belong to my var. *ramosa*, but I do not think that this can be at all sharply distinguished from the type.

One of the specimens contains a number of grapnel spicules, which seem to be foreign, as they were seen only in one small part.

Previously known Distribution. Gulf of Mannar, Ceylon Seas (Dendy) ; Aru Islands, Arafura Sea (Hentschel).

Register Numbers, Locality, &c. XX. 5, 10, Adatra.

42. **Echinodictyum gorgonioides** n. sp.—(Plate I., Figs. 7*a*–7*b'* ; Plate IV., Fig. 23.)

So far as external appearance goes this is perhaps the prettiest sponge in the collection, with an extremely characteristic mode of growth. There are three separate pieces of it, but they all come from the same locality and may possibly be parts of one and the same specimen. The sponge (Fig. 23) consists of a number of slender stems springing close together from a somewhat spreading, encrusting base. The stems rise almost vertically from the points of attachment and bifurcate repeatedly and very frequently. All the branches—and no distinction can be drawn between stems and branches—are of about the same diameter, say 3 mm. They mostly lie in approximately the same plane and occasionally anastomose with one another. The surfaces of the branches are subrugose and minutely conulose. The branches terminate in rounded extremities and there are no conspicuous vents. The whole growth reminds one forcibly of some species of Gorgonia.

The total height of the largest piece is about 110 mm., and the greatest width nearly as much.

The colour in spirit (after formalin) is very pale brown. The texture is fairly tough, very compressible and resilient.

The skeleton is composed of stout primary fibres, which branch repeatedly as they approach the surface and are more or less interconnected by secondary fibres. The main fibres consist chiefly of plumose columns of spined styli, some of which are completely embedded in the pale-coloured spongin, while others echinate the surface of the fibre at various angles. The secondary fibres consist mainly or entirely of spongin, echinated more or less by the spined styli.

Fairly numerous oxea (tornotoxea) accompany the fibres or are scattered between them, but they rarely, if ever, form a spicular core to the fibre. They are not nearly so numerous as the spined styli, which form the chief part of the spiculation.

Spicules. (1) Oxea (tornotoxea) (Fig. 7*a*) ; straight, slender, thicker at one end than at the other, abruptly pointed at the thicker end, gradually sharp-pointed at the other ; occasionally with a bulbous inflation at from $\frac{1}{3}$ to $\frac{1}{4}$ the length from the thicker end ; size about 0·14 by 0·0041 mm.   (2) Spined styli (acanthostyli) (Figs. 7*b*, 7*b*') ; tapering gradually from the base to the sharply pointed apex ; straight ; rather sparsely covered with small spines, except towards the apex ; size variable, say about 0·11 by 0·01 mm. ; a number of very slender forms also occur (Fig. 7*b*').

The sponge also contains numerous other types of spicule, most of which, at any rate, have certainly been derived from other sponges, including Spirastrella, Esperella, Donatia and Reniera.   There are, however, a fair number of stoutish, strongly angulated toxa, a very few minute Clathria-like, palmate isochelæ, and a fair number of long, straight, slender tylostyles with well-developed heads, which may possibly belong to the species.

The plumose character of the main skeleton fibres suggests a close affinity with the genus Plumohalichondria.

Register Number, Locality, &c.   XXI. 1, 3, 4–7 fms. off Rupan Bandar and Kutchegudh, 8.12.05.

43. Raspailia fruticosa var. tenuiramosa Dendy.

Raspailia fruticosa Dendy [1887].

Raspailia fruticosa var. *tenuiramosa* Dendy [1905].

Raspailia fruticosa var. *aruensis* Hentschel [1912].

The two specimens in the collection differ from the Ceylonese types of the variety in their much more sparingly branched character, being, in fact, less instead of more bushy than the type of the species.   Both specimens, however, have much more slender branches than the type of the species.   As regards skeleton arrangement and spiculation they agree very closely with the types of the variety.

Previously known Distribution of the Species.   Gulf of Mannar (Dendy) ; Aru Islands, Arafura Sea (var. *aruensis* Hentschel).

Register Numbers, Localities, &c.   II. 13, off Poshetra, January 7, 1906 ; XXIX. 2, N. of Poshetra, 3¾–4 fms., 20.12.05.

44. Acarnus tortilis Topsent [1892 *bis*, 1897, 1904].

I identify with this widely distributed but apparently rare species a small, presumably encrusting sponge of a brown colour and irregular shape.   The spiculation agrees closely with that described and figured by Topsent for his species, even down to the minute spination of the bases of the stout styli and the ends of the diactinal megascleres.   The very characteristic grapnel-spicules, with usually four strongly recurved, sharp hooks at the apex, strongly spined shaft, and base with spines curved in the opposite direction, are identical.   The toxa, however, appear to be all of the strongly arcuate form, though varying much in dimensions.   The chief difference that I have been able to detect, however, lies in the presence of a considerable amount of spongin, partially uniting and enveloping some of the spicules, but this difference can hardly be regarded as of specific value and may be associated with temperature conditions. A great many foreign spicules are present in the sponge, possibly owing to its association with other sponges in the dredge.

Previously known Distribution. Mediterranean, Azores, Amboina (Topsent).

Register Number, Locality, &c. XXIX. 3, N. of Poshetra, 3¾–4 fms., 20.12.05.

45. Bubaris radiata n. sp. (Plate I., Figs. 8a–8b ; Plate IV., Figs. 24a, 24b).

Sponge (Figs. 24a, 24b) encrusting, cushion-shaped, composed of close-set, stout, radiating skeletal columns, united by a small quantity of gelatinous soft tissue. Surface conulose owing to the projecting ends of the spicular columns, with a thin, translucent dermal membrane in the grooves between the conuli. No oscula or pores seen. The largest piece (R.N. II. 10) measures about 18 mm. in maximum diameter and 6 mm. in thickness, the thickness being equal to the length of the spicular columns, which extend from the base to the upper surface almost without branching and approximately parallel to one another. The individual columns are about 1 mm. in diameter. The colour in spirit is light, dull yellow.

A second specimen (R.N. III. 7 a), rather smaller, looks like a fragment of an almost spherical sponge. The spicular columns are rather stouter, a little more branched, and radiate almost from a common centre. The colour in spirit is rather darker.

The skeleton consists of very strong columns of spicules arranged in a plumose fashion. Each column contains a very dense axis composed of an interlacement of short strongyla, in which are implanted the bases of some of the stout styli which radiate obliquely outwards and upwards from the axis. There is sometimes a tendency to the formation of secondary plumose spicular columns coming off from the axis of the primary. At the ends of the columns the apices of the styli project freely beyond the surface of the sponge, but lower down between the columns they are completely enveloped in gelatinous tissue.

Spicules. (1) Stout styli (Fig. 8a), more or less bent towards the base, tapering gradually to a sharply pointed apex ; commonly measuring about 0·55 by 0·026 mm., often smaller and occasionally a trifle larger. (2) A few very much longer and more slender styli (Fig. 8a‴) occur in boiled-out preparations, measuring up to about 1·3 by 0·016 mm. (3) Strongyla (Fig. 8b) ; comparatively short and irregularly curved or bent to a varying but not very high degree, approximately equal-ended ; size commonly about 0·26 by 0·01 mm., but variable. Intermediate forms are represented by Figs. 8a′ and 8a″.

In general appearance and skeleton arrangement this sponge reminds one very strongly of the common Ceylon species *Aulospongus tubulatus*. The latter, however, has not got any strongyla in the axes of the plumose spicular columns, while, on the other hand, it has got minutely spined, echinating styli, which are absent in the present species.

From *Bubaris vermiculata* the species differs in its mode of growth and in the replacement of the " vermicular " spicules by much less bent strongyla.

The descriptions, measurements and figures of the spicules, as well as the figures of the external form, are taken from R.N. III. 7 *a*, which may be regarded as the type of the species.

Register Numbers, Localities, &c. II. 10, off Poshetra, 7 January, '06 ; III. 7*a*, *b*, (two fragments), off Dwarka.

46. **Spirastrella vagabunda** var. **tubulodigitata** Dendy.—(Plate IV., Fig. 25).
(For possible Synonymy, *vide* Vosmaer [1911]).

There is in the collection a single specimen (Fig. 25) which agrees very closely in external form and spiculation with the Ceylon types. It consists of a single tubular process with terminal vent, and a few much smaller, irregularly ramified, blind processes, all arising from a common base containing much coarse sand.

Previously known Distribution of the Variety. Gulf of Mannar and Ceylon Seas (Dendy).

Register Number, Locality, &c. IV. 10, off S.W. coast of Beyt Island.

47. **Placospongia carinata** (Bowerbank).
(For Literature and Synonymy *vide* Vosmaer and Vernhout [1902] and Dendy [1905]).

This remarkable sponge appears to be quite common in the Indian Ocean.

Previously known Distribution. Tropical Seas between 30° N. and 20° S. of the equator (Vosmaer and Vernhout, &c.).

Register Numbers, Localities, &c. XXVI. 9, Adatra Reefs, 25 December, '05 ; XXVII. 2, 3 and XXXIII. 4, Dhed Mora and adjacent rocky ground between Beyt and Aramra, 1 fm., 21.12.05.

48. **Cliona coronaria** (Carter).
Suberites coronarius Carter [1882, 1887].

This interesting species was first described by Carter from specimens in the Bowerbank collection in the British Museum, coming from Honduras, Jamaica and the Bahamas. These specimens were described as " massive, lobate, verrucose on the surface." The characteristic microsclere was described as a " Spinispirula consisting of one bend, semicircular, with the spines on the outside and over the ends only ; spines capitate and in single file."

Mr. Carter's original preparations of these three sponges are in my possession, and I am able to verify the general accuracy of his brief description and figures of the spiculation. The capitate character of the spines of the microscleres is, however, not always recognisable, and may be in part due to optical illusion.

In 1887 the same species was recorded by Mr. Carter from the Mergui Archipelago and some interesting particulars added as to its mode of growth, as follows :—" Its growth is more remarkable than in that [the Honduras] example, for it is laminar, and extends in a horizontal direction for several square inches ; the superficial stratum, which is comparatively thin and buff-yellow in colour, changes to black or dark brown in the cancellated cavities to be presently mentioned for half an inch downwards, where it rests on granite. The explanation of this abrupt termination is that the lower portion is mingled with a layer of coral which has been cancellated by the excavating habit of these sponges, which exhibit an apparent fondness for calcareous material, whether in a mineral or organic form."

Mr. Carter thus clearly recognises that his *Suberites coronarius* may, at any rate under some circumstances, be an excavating sponge. A specimen in Mr. Hornell's collection (R.N. XXVI. 10), also excavating and encrusting a piece of coral, agrees very closely with the Mergui specimen, and its examination, I think, fully justifies the transference of the species to the genus Cliona. In all probability the massive West Indian specimens stand in exactly the same relation to the excavating Indian Ocean specimens as does the massive " *Raphyrus griffithsii* " to the excavating *Cliona celata* of European seas, or to the encrusting form of the latter described by Topsent [1900].

Mr. Hornell's specimen consists, in the first place, of a thin crust (about 1 mm. thick), with smooth outer surface. The underlying coral has been eroded and largely disintegrated by the sponge, giving rise to the " cancellated " structure described by Mr. Carter, which, when teased up and examined microscopically, is seen to consist of a mixture of coral fragments and sponge.

In the outer portion of the sponge the skeleton consists of dense, irregular wisps of tylostyles running towards the surface, where they form a thick dermal pile with outwardly directed apices. In the cancellated portion the tylostyles seem to be quite irregularly scattered. The spiculation consists of tylostyles and spirasters only ; I have seen no oxea such as sometimes occur in Cliona (*fide* Topsent). The tylostyles are straight or very slightly curved, usually very sharply and gradually pointed. They have well developed, almost spherical heads. When fully grown they measure about 0·35 by 0·012 mm., with head 0·014 mm. in diameter. The greatest diameter of the shaft is at about one-third of the distance from head to apex. The spirasters are very slender and measure about 0·02 mm. in a straight line from end to end. Many of them have the typical " semicircular " form described by Carter, but many show a very obvious spiral twist. They resemble bent fragments of broken fretsaws. It is not easy to assure oneself that the short spines are really capitate, though they may have the appearance of being so.

As might be expected, the spiculation, as well as the mode of growth, of the Okhamandal sponge seems to agree more closely with that of the Mergui than with that of the West Indian specimens. Thus the heads of the tylostyles are more nearly spherical in both the Indian Ocean specimens (compare Carter's figures of the Mergui and West Indian sponges), but I do not think there is anything to justify a specific distinction.

It should be noted that the Okhamandal specimen was first preserved in formalin, which may possibly have assisted in bringing about the eroded condition of the coral on which it is growing.

Some larger fragments from the same locality (R.N. XXVI. 2), possibly parts of the same specimen, have been completely stripped off from the substratum in the form of a thin sheet about 48 square centimetres in extent and about 1 mm. thick. They have an almost smooth surface, minutely reticulate in places when viewed under a lens, and no visible vents.

Previously known Distribution. Honduras, Jamaica, Bahamas, Mergui (Carter).

Register Number, Locality, &c. XXVI. 2, 10, Adatra Reefs, 25 December, '05·

49. Suberites carnosus (Johnston) var.

(For Literature and Synonymy, *vide* Topsent [1900]).

This common and widely distributed species is represented in the collection by two specimens of massive, subovoid form, without any indication of stalks, and with no visible vents. One of the two specimens (R.N. XIX. 3) measures about 44 by 29 mm. and is compact and solid. The other (R.N. XIX. 2), of about the same size, has been damaged in collecting, and a superficial cortical layer, made coherent by the extremely dense surface-skeleton, has been separated in large measure from the remainder, which seems to have shrunk away or been partially removed, so that the specimen looks like a broken egg with a dried up yolk adherent to the shell at one side only. In both the surface is smooth but very minutely velvety, and the colour throughout (in spirit) is pale yellow.

The skeleton consists exclusively of tylostyles. In the interior of the sponge these spicules are thickly but quite irregularly scattered, while at the surface they arrange themselves with their apices projecting outwards, not in distinct brushes but in a dense, continuous layer.

The tylostyles are approximately straight, gradually and sharply pointed, and with well rounded heads only occasionally of the " enormispinulate " type. They vary in size up to about 0·77 by 0·018 mm.

A direct comparison with a preparation of Johnston's type-specimen shows that the latter has decidedly smaller spicules with a strong tendency to develop a secondary inflation just below the head. The measurements of the spicules in our variety also exceed the range of variation for the species given by Topsent.

Previously known Distribution. Atlantic, Mediterranean, Red Sea, Indian Ocean and Australian Seas (*vide* Ridley and Dendy [1887] and Topsent [1900]).

Register Numbers, Locality, &c. XIX. 2, 3, Vamiani Point, January 5, '06·

50. Suberites flabellatus Carter.

Suberites flabellatus Carter [1886].

? *Suberites globosa* (elongated form) Carter [1886].

? *Suberites carnosus* Keller [1891].

Suberites flabellatus Dendy [1897].

A single good specimen in the collection closely resembles in external form Keller's figure of "*Suberites carnosus* " from the Red Sea. It agrees very closely, however, with the Victorian species described by Carter and myself, and I have no hesitation in identifying it with that, which may be, after all, merely a variety of *S. carnosus*. The tylostyles in the Okhamandal sponge are long, straight and slender, and they all seem to have well-developed heads, usually of the "enormispinulate " type. In the interior of the sponge they attain a length of about 0·7 mm., with a diameter of about 0·01 mm.; but they are much smaller in the surface-brushes.

Previously known Distribution. Near Port Phillip Heads, Australia (Carter, Dendy); ? Red Sea (Keller).

Register Number, Locality, &c. II. 1, off Poshetra, January 7, '06·

51. Suberites cruciatus Dendy [1905].

I identify with this species, originally described from Ceylon, two small, irregularly massive specimens, each of which shows a tendency to give off digitiform processes. The larger of the two measures about 18 mm. in greatest length by 9 mm. in greatest breadth. The colour in spirit is pale yellow. The surface is uneven but nearly smooth, and shows small, rounded, translucent (pore ?—) areas, as described for the type.

The arrangement of the skeleton agrees on the .whole with that found in the type. The tylostyles exhibit the same peculiar form of the heads but are considerably larger than in the type ; in fact, they may be nearly twice as large, at any rate in the deeper parts of the sponge, while they diminish in size in the surface-brushes. A large number of tylostyles occur scattered tangentially in the dermal membrane between the ends of the surface-brushes, a feature which is not conspicuous in the type.

Previously known Distribution. Ceylon (Dendy).

Register Numbers, Locality, &c. III. 10 *a* and *b* (possibly parts of same specimen) ; dredged off Dwarka.

52. Polymastia gemmipara n. sp.—(Plate I., Figs. 9a, 9b ; Plate IV., Figs. 26a, 26b).

The single specimen (Figs. 26a, 26b) has evidently been torn off from the substratum, part of which, in the form of a small pebble, remains attached at one side. It now has the form of a hollow, thin-walled sac, widely open below, where it has been damaged, and produced above into five slender, hollow processes or fistulæ, one of which is branched. In life, no doubt, the interior of the sponge was filled with soft, pulpy, choanosomal tissue, part of which remains in the form of irregular masses adherent to the inner surface of the cortex, as shown at *ch.* in Fig. 26*b*.

The thin cortex is supported internally by a few very stout spicular columns, one of which is shown at *sp. c.* in the illustrations. No doubt in life these columns, of which I have only been able to find two in the specimen before me, were attached to the substratum, and formed pillars supporting the dome-like cortex.

The fistulæ are very remarkable structures. They have no visible openings, but most of them taper off distally into fine threads, which in three cases are swollen out to form either one or two small buds, as shown at *b.* in the figures.

The colour of the cortex (in alcohol) is white, that of the soft internal pulp very pale yellow.

The greatest diameter of the specimen, at the base, is 17 mm.

The skeleton may be subdivided as follows : (1) that of the soft internal choanosome or pulp consists of loosely scattered tylostyles or subtylostyles, varying in size but mostly small and often very slender ; (2) that of the internal spicular columns consists of a dense mass of relatively large subtylostyles closely packed together lengthwise ; (3) that of the cortex consists of an inner and an outer portion ; the inner portion is a fairly close interlacement of the larger subtylostyles lying tangentially, not much more than one layer thick, with a tendency to the formation of broad bands, which can be seen under a pocket-lens converging towards the bases of the fistular processes ; the outer portion is a thin but rather close pile or fur of short tylostyles or subtylostyles arranged more or less vertically to the surface, with outwardly projecting apices ; · (4) that of the fistular processes is merely a continuation of the cortical skeleton. The principal bundles of the larger subtylostyles run lengthwise in the wall of the fistula ; where the fistula is drawn out into a solid thread at the end they unite to form a single stout spicular fibre, and in this region the surface-fur of short spicules is almost absent, but becomes strongly developed again when the thread swells out to form a bud.

Spicules. (1) Subtylostyles of the internal columns and cortex (Fig. 9*a*) ; straight, slender, tapering gradually to a finely pointed apex, also tapering gradually to the base, where there is usually a very slightly developed head ; the base sometimes appears constricted somewhat suddenly, the " head " being a good deal narrower than the adjacent part of the shaft, as in the specimen figured ; commonly measuring about 0·72 by 0·0164 mm. ; (2) small tylostyles (Fig. 9*b*) ; chiefly in the surface pile ; usually curved ; apex gradually sharp-pointed ; with rather feebly developed head of smaller diameter than the middle of the spicule ; size about 0·15 by 0·0068 mm. ; (3) spicules of the soft internal pulp ; a mixture of the two kinds already described, together with intermediate forms and very slender forms which are probably young ; perhaps never quite so large as in the spicular columns and cortex, and usually much smaller.

This species seems to be intermediate in character between the genera Polymastia and Quasillina as at present understood, the soft, pulpy internal structure and thin cortex being suggestive of the latter, and the hollow fistular processes of the former. It seems quite possible that, after all, the distinction between these two genera may have to be abandoned.[1]

The process of bud-formation by gemmation from the filiform ends of tapering fistulæ appears to be identical with that described, with admirable illustrations, by Merejkowsky [1878] in *Polymastia (Rinalda) arctica*, a species from the White Sea. A somewhat similar process is, of course, familiar in certain other genera, such as Donatia.

In *Polymastia arctica* there are one or more osculum-bearing fistulæ between the gemmiparous ones. This may also be the case with the present species, but the specimen is not in a fit condition to enable me to determine the point with certainty.

Register Number, Locality, &c. IV. 7, dredged off S.W. coast of Beyt Island.

53. **Megalopastas retiaria** n. sp.—(Plate IV., Fig. 27.)

This is certainly one of the most interesting, and at the same time one of the most beautiful species in the collection, in which it is well represented by three good spirit specimens and a couple of washed-out skeletons. The general appearance of the sponge is well shown in Fig. 27, which represents the best preserved specimen twice the natural size. The sponge is irregularly lobose, or simply massive. Apparently it is sessile, with several points of attachment to relatively small foreign objects, as though it had grown on a gravelly or shelly bottom. The surface is produced into moderate-sized, acute conuli, scattered at rather wide and irregular intervals, the height of the conuli being about 1–2 mm. Internally the sponge is cavernous and the wide vestibules open on the surface by large pseudoscula. These pseudoscula, of various shapes and sizes, are frequently (perhaps always in life) covered over by a very beautiful network, which is a continuation of the delicately reticulate dermal membrane, but with much coarser meshes. The dermal membrane in general covers the whole sponge like a gauzy veil. It is strengthened by a close reticulation of delicate, deeply staining, fibrillar bands (not horny). Under a pocket-lens the principal lines of this network appear (in the spirit specimens) as fine white lines radiating from the apices of the conuli into the hollows between, where they lose themselves in a network of finer lines. The meshes of this network are occupied by the pore-bearing dermal membrane, which is itself reduced to a secondary, quite microscopic network by the very numerous inhalant pores, each only about 0·03 mm. in diameter. The dermal membrane is interrupted here and there, pretty frequently, by small circular apertures about 1 mm. in diameter. These apertures have well-defined margins, being bounded each by an unusually large, circular

---

[1] *Cf.* my remarks on *Quasillina brevis* in *Journ. Linn. Soc. Zool.*, Vol. XXXII. (1914), p. 271.

mesh of the network of fibrillar bands described above. They are the openings of deep cylindrical canals of similar diameter, and probably represent true oscula. Between them numerous smaller apertures, the openings of the inhalant canals leading from the subdermal cavities, can be seen through the transparent dermal membrane. In some places the soft tissues seem to have shrunk away from the skeleton, showing the coarsely reticulate, dark-coloured, horny fibre projecting beyond the surface ; while the pale-coloured growing points of the primary fibres appear in the conuli, supporting the dermal membrane.

The largest specimen (R.N. VIII.) is an irregular lobose mass about 100 mm. in diameter. The texture (in spirit) is very soft and compressible, but resilient owing to the coarse, stiff, horny skeleton. The colour in life (R.N. VIII.) was recorded as pink, in spirit all three specimens are very pale grey.

Of two specimens preserved in formalin (R.N. XXIX. 1) practically nothing remained but the skeleton.

The skeleton is a very coarse reticulation of dark-brown, almost black, horny fibre. It is only occasionally possible to recognise main stems springing from points of attachment. Indeed the abundant development of secondary connecting fibres almost completely conceals any original tree-like growth there may have been, except towards the surface, where branching main fibres terminate in the conuli. The connecting fibres are developed chiefly in the angles of the main branching system, with the result that we can distinguish meshes of two quite distinct orders, large and small, the small ones occurring in groups at the nodes of the reticulation formed by the large ones. The large meshes average perhaps 5 mm. in diameter, the small ones, say, about 0·5 mm., but very variable. The thickest primary fibres measure up to about 0·34 mm. in diameter and the thinnest secondary ones about 0·034 mm., between which extremes all gradations occur. The fibres are entirely free from foreign inclusions and consist of more or less numerous concentric layers of spongin. It is not possible to distinguish sharply between the so-called pith and the surrounding spongin-lamellæ, and, in spite of the numerous descriptions in which such distinction is insisted upon, I believe that this is frequently the case in the Aplysillidæ.

There is no dermal or subdermal reticulation of horny fibres.

The canal-system is typically aplysillid. The flagellate chambers are very large and thimble-shaped, measuring about 0·17 by 0·085 mm. Favourable sections show them arranged in a single, much folded layer between the inhalant and exhalant canals. Each chamber has numerous prosopyles.

Register Numbers, Localities, &c. II. 3, off Poshetra, January 7, '06 ; IV. 13, off S.W. coast of Beyt Island ; VIII., Mangunda Reef ; XXIX. 1 (two skeletons), N. of Poshetra, 3¾–4 fms., 20.12.05.

54. **Darwinella australiensis** Carter.
Darwinella australiensis Carter [1885].
Darwinella australiensis Lendenfeld [1889].
Darwinella australiensis Topsent [1905].
Darwinella australiensis Hentschel [1912].

There are in the collection the remains of two or three specimens of this sponge, which has thickly encrusted some of the large polychæte worm-tubes before mentioned. Unfortunately the specimens were preserved in formalin and are very badly macerated. There can, however, be little doubt about the identification.

The surface, where preserved, is strongly conulose, the colour (in alcohol after formalin) is dull red. The largest specimen is about 25 mm. in maximum thickness (the height of the largest skeletal trees).

The skeleton is composed of elegant little trees of pale amber-coloured horny fibre, growing up vertically and side by side from a thin basal layer of spongin. The trees are well branched, but anastomosis between the branches takes place only rarely.

The horny spicules seem to be much less numerous than in Australian specimens. They are triradiate, with long, slender, gradually sharp-pointed rays. They all seem to lie freely in the soft tissues. I have measured the rays up to about 0·85 mm. in length. I have observed in some of the older fibres apparently the same parasitic fungus (?) as I described [1905] in the case of *Darwinella simplex*, Topsent, from Ceylon. It now appears to me very doubtful whether that species is distinguishable from *D. australiensis*.

Previously known Distribution. Victoria, Australia (Carter, Lendenfeld) ; Arafura Sea (Hentschel).

Register Number, Locality, &c. XXV. 2, Kiu, littoral at low water, 24.12.05.

55. **Spongelia fragilis** var. **ramosa** (Schulze).
Spongelia pallescens subspecies *fragilis* var. *ramosa* Schulze [1879].
Spongelia fragilis var. *irregularis* (pars) Lendenfeld [1889].
Spongelia fragilis var. *ramosa* Dendy [1905].
? *Spongelia fragilis* var. *clathrata* Hentschel [1912].

There are several specimens of this sponge in the collection. I pointed out in my Ceylon Report that the species, at any rate, is probably cosmopolitan, if not the variety.

Previously known Distribution of Variety. Adriatic (Schulze) ; Ceylon (Lendenfeld, Dendy) ; ? Aru Islands (Hentschel).

Register Numbers, Localities, &c. XII., XXXII. 3, off Beyt ; XXV. 5, Kiu, littoral.

56. **Spongelia cinerea** (Keller).

Dysidea cinerea Keller [1889].

Spongelia elastica var. *crassa* Dendy [1905].

There are three specimens of this sponge in the collection, one large one (R.N. IV. 1) Very similar to that figured in my Report on Professor Herdman's Ceylon Sponges, and two small ones more nearly resembling the figure given by Keller. I think there can be little doubt that my var. *crassa* is identical with Keller's *Dysidea cinerea* from the Red Sea, which is possibly merely a variety of *Spongelia elastica*, which, in turn, according to Schulze [1879], is merely a subspecies of *Spongelia pallescens*. It is very doubtful how far it is desirable to separate these different forms.

Previously known Distribution. Red Sea (Keller) ; Gulf of Mannar (Dendy). .

Register Numbers, Localities, &c. IV. 1, 15, 23, dredged off S.W. coast of Beyt Island.

57. **Spongelia elegans** Nardo var.

(For Synonymy, *vide* Schulze [1879] and Lendenfeld [1889].)

A number of branched and anastomosing, digitiform pieces, all from the same locality and probably all parts of one specimen, closely resemble Schulze's figure of *Spongelia elegans*. The specimen seems, however, to have been less robust than that figured by Schulze and the conuli less strongly developed. The colour in spirit (after formalin) is light brown ; texture very soft and compressible. The skeleton is a loose, irregular network of rather slender, strongly arenaceous fibre, in which the fibres are all equally charged with sand, there being no secondary horny fibres more or less free from sand and connecting the fibres which run into the conuli, as described by Schulze. Stained preparations, however, show darkly staining fibrillar bands of nucleated cells running lengthwise through the sponge. These commonly contain a few sand-grains. Their relation to the ordinary skeletal fibres is doubtful, though they appear to be connected with them in places.

Schulze says nothing about the occurrence and distribution of spongin, and one merely assumes that the sand-free secondary fibres in his material are composed of that substance. In the Okhamandal specimen spongin is nowhere strongly developed, and I have recognised it only in the sandy fibre.

Previously known Distribution. Mediterranean (Nardo, Schmidt, Schulze, &c.) ; East Coast of Australia (Lendenfeld).

Register Number, Locality, &c. XXIII. 4, off Dwarka, 15–17 fms., 12.12.05.

58. **Hippospongia clathrata** (Carter).

Hircinia clathrata Carter [1881].
Hircinia clathrata Dendy [1887, 1889].
Hyatella clathrata Lendenfeld [1889].
Hippospongia clathrata Dendy [1905].

There are several large pieces of this sponge in the collection.

Previously known Distribution. Gulf of Mannar and Red Sea (Carter); Gulf of Mannar (Dendy); Australia (Dendy, Lendenfeld); American coast of N. Atlantic (Lendenfeld).

Register Numbers, Localities, &c. XVIII. 1, channel, W. of Beyt Island, 3–4 fms., January, '06 ; XXXII. 6, off Beyt.

LIST OF LITERATURE REFERRED TO.

1915. Annandale, N. " Fauna of the Chilka Lake. Sponges." (*Mem. Indian Museum.*. Vol. V.).

1903. Arnesen, Emily. " Spongien von der norwegischen Küste. II. Monaxonida : Halichondrina." (*Bergens Mus. Aarbog*, 1903, No. 1.)

1905. Baer, L. "Silicispongien von Sansibar, Kapstadt und Papeete." (*Archiv für Naturgeschichte*, Jahrgang 72, Bd. 1).

1866. Bowerbank, J. S. " A Monograph of the British Spongiadæ." Vol. II.

1873. Bowerbank, J. S. " Report on a Collection of Sponges found at Ceylon by E. W. H. Holdsworth, Esq." (*Proc. Zool. Soc. Lond.*, 1873.)

1875. Bowerbank, J. S. " Contributions to a General History of the Spongiadæ." Part VII. (*Proc. Zool. Soc. Lond.*, 1875.)

1869. Carter, H. J. " Description of a Siliceous Sand Sponge found on the South-East Coast of Arabia (*Tethya dactyloidea*)." (*Ann. and Mag. Nat. Hist.*, Series 4, Vol. III.)

1872. Carter, H. J. " Additional Information on the Structure of *Tethya dactyloidea*, Carter." (*Ann. and Mag. Nat. Hist.*, Series 4, Vol. IX.)

1873. Carter, H. J. " On Two New Species of Gummineæ (*Corticium abyssi, Chondrilla australiensis*) with special and general Observations." (*Ann. and Mag. Nat. Hist.*, Series 4, Vol. XII.)

1874. Carter, H. J. " Descriptions and Figures of Deep-sea Sponges and their Spicules from the Atlantic Ocean," &c. (*Ann. and Mag. Nat. Hist.*, Series 4, Vol. XIV., pp. 207 and 245.)

1879. Carter, H. J. " Contributions to Our Knowledge of the Spongida." (*Ann. and Mag. Nat. Hist.*, Series 5, Vol. III., pp. 284 and 343.)

1880. Carter, H. J. " Report on Specimens Dredged up from the Gulf of Manaar," &c. (*Ann. and Mag. Nat. Hist.*, Series 5, Vol. VI., pp. 35 and 129).

1881. Carter, H. J. " Supplementary Report on Specimens Dredged up from the Gulf of Manaar," &c. (*Ann. and Mag. Nat. Hist.*, Series 5, Vol. VII.)

K

1882.   Carter, H. J. " Some Sponges from the West Indies and Acapulco," &c. (*Ann. and Mag. Nat. Hist.*, Series 5, Vol. IX., pp. 266 and 346.)

1882 *bis.* Carter, H. J. " New Sponges, Observations on Old Ones, and a Proposed New Group (Phœodictyina)." (*Ann. and Mag. Nat. Hist.*, Series 5, Vol. X.)

1885.   Carter, H. J. " Descriptions of Sponges from the Neighbourhood of Port Phillip Heads, South Australia." (*Ann. and Mag. Nat. Hist.*, Series 5, Vol. XV., p. 196.)

1885. *bis.* Carter, H. J. " Descriptions of Sponges from the Neighbourhood of Port Phillip Heads, South Australia, continued." (*Ann. and Mag. Nat. Hist.*, Series 5, Vol. XVI., p. 277.)

1886.   Carter, H. J. " Descriptions of Sponges from the Neighbourhood of Port Phillip Heads, South Australia, continued." (*Ann. and Mag. Nat. Hist.*, Series 5, Vol. XVII., p. 112.)

1887.   Carter, H. J. " Report on the Marine Sponges, chiefly from King Island, in the Mergui Archipelago," &c. (*Journ. Linn. Soc. Zool.* Vol. XXI.).

1887. *bis.* Carter, H. J. " Description of *Chondrosia spurca*, n. sp., from the South Coast of Australia." (*Ann. and Mag. Nat. Hist.*, Series 5, Vol. XIX.)

1887.   Dendy, A. " The Sponge Fauna of Madras," &c. (*Ann. and Mag. Nat. Hist.*, Series 5, Vol. XX.)

1889.   Dendy, A. " Report on a Second Collection of Sponges from the Gulf of Manaar." (*Ann. and Mag. Nat. Hist.*, Series 6, Vol. III.)

1890.   Dendy, A. " Observations on the West Indian Chalinine Sponges, with Descriptions of New Species." (*Trans. Zool. Soc. Lond.* Vol. XII.)

1895.   Dendy, A. " Catalogue of Non-Calcareous Sponges collected by J. Bracebridge Wilson," &c. Part 1. (*Proc. Royal, Soc. Victoria.* Vol. VII. n. s.)

1896.   Dendy, A. " Catalogue of Non-Calcareous Sponges collected by J. Bracebridge Wilson," &c. Part 2. (*Proc. Royal Soc. Victoria.* Vol. VIII. n. s.)

1897.   Dendy, A. " Catalogue of Non-Calcareous Sponges collected by J. Bracebridge Wilson," &c. Part 3. (*Proc. Royal Soc. Victoria.* Vol. IX. n. s.)

1905.   Dendy, A. " Report on the Sponges collected by Professor Herdman at Ceylon in 1902." (*Report on Pearl Oyster Fisheries.* Part 3. *Royal Society.*)

1915.   Dendy, A. " Report on the Calcareous Sponges," &c. (*Report to the Government of Baroda on the Marine Zoology of Okhamandal in Kattiawar*, Part II.)

1916.   Dendy, A. " Report on the Homosclerophora and Astrotetraxonida collected by H.M.S. *Sealark* in the Indian Ocean." (*Trans. Linn. Soc. Lond. Zool.* Series 2, Vol. XVII. In the Press.)

1870.   Ehlers, E. " Die Esper'schen Spongien in der Zoologischen Sammlung der K. Universität Erlangen." (*Programm zum Eintritt in den Senat, Erlangen*, 1870.)

1794–1806. Esper, E. J. C. " Fortsetzung der Pflanzenthiere." (*Nürnberg.*)

1867.   Gray, J. E. " Notes on the Arrangement of Sponges, with the Descriptions of some New Genera." (*Proc. Zool. Soc. Lond.*, 1867.)

1909.   Hentschel, E. " Tetraxonida " 1 Teil. (*Die Fauna Südwest-Australiens.* Bd. II. Jena.)

1911.   Hentschel, E. " Tetraxonida " 2 Teil. (*Die Fauna Südwest-Australiens.* Bd. III. Jena.)

1912.   Hentschel, E. " Kiesel- und Hornschwämme der Aru-und Kei-Inseln." (*Abh. Senckenberg. Nat. Ges. Frankfurt a.M.* Bd. XXXIV.)

1914 (?) Hentschel, E. " Monaxone Kieselschwämme und Hornschwämme." (*Deutsche Südpolar-Expedition* 1901–1903.)

1880.   Keller, C. " Neue Cœlenteraten aus dem Golf von Neapel." (*Archiv Mikrosk. Anat.* Vol. XVIII.)

1889.   Keller, C. " Die Spongienfauna des rothen Meeres." I. Hälfte. (*Zeit. wiss. Zool.* Bd. XLVIII.)

1891.   Keller, C. " Die Spongienfauna des rothen Meeres." II. Hälfte. (*Zeit. wiss. Zool.* Bd. LII.)

1900.   Kirkpatrick, R. " On the Sponges of Christmas Island." (*Proc. Zool. Soc. Lond.*, 1900.)

1914.   Lebwohl, F. " Japanische Tetraxonida." I., II., III., IV. (*Journ. Coll. Sci. Imp. Univ. Tokyo.* Vol. XXXV.)

1886.   Lendenfeld, R. von. " A Monograph of the Australian Sponges. Part IV. The Myxospongiæ." (*Proc. Linn. Soc. N.S.W.* Vol. X.)

1889.   Lendenfeld, R. von. " A Monograph of the Horny Sponges." (*Royal Society.*)

1897.   Lendenfeld, R. von. " Spongien von Sansibar." (*Abh. Senckenberg. nat. Ges. Frankfurt a.M.* Bd. XXI.)

1903.   Lendenfeld, R. von. " Tetraxonia." (*Das Thierreich, Berlin.*)

1906.   Lendenfeld, R. von. " Die Tetraxonia." (*Wiss. Ergeb. d. deutsch. Tiefsee-Expedition auf d. Dampfer " Valdivia,"* 1898–1899. Jena.)

1910.   Lendenfeld, R. von. " The Geodidæ." (*" Albatross " Expedition. Mem. Mus. Comp. Zool. Harvard College.* Vol. XLI.)

1898.   Lindgren, N. G. " Beitrag zur Kenntniss der Spongienfauna des Malayischen Archipels und der chinesischen Meere." (*Zool. Jahrb. Ab. Syst.* Bd. XI.)

1878.   Merejkowsky, C. " Études sur les Éponges de la Mer Blanche." (*Mém. Acad. St. Petersburg.* Vol. XXVI.)

1884.   Ridley, S. O. " Spongiida." (*Zool. Coll. H.M.S. " Alert." Brit. Mus. Nat. Hist.*)

1887.   Ridley, S. O. and Dendy, A. " Report on the Monaxonida," &c. (" *Challenger* " *Reports Zoology.* Vol. XX.)

1911.   Row, R. W. H. " Report on the Sponges collected by Mr. Cyril Crossland in 1904–5. Part II. Non-Calcarea." (*Reports on the Marine Biology of the Sudanese Red Sea. Journ. Linn. Soc. Lond. Zool.* Vol. XXXI.)

1862.   Schmidt, O. " Die Spongien des adriatischen Meeres." (*Leipsig.*)

1877.   Schulze, F. E. " Untersuchungen über den Bau und die Entwicklung der Spongien. III. Die Familie der Chondrosidæ." (*Zeit. wiss. Zool.* Bd. XXIX.)

1879.   Schulze, F. E. " Untersuchungen über den Bau und die Entwicklung der Spongien. VI. Die Gattung Spongelia." (*Zeit. wiss. Zool.* Bd. XXXII.)

1888.   Sollas, W. " Report on the Tetractinellida," &c. (" *Challenger* " *Reports. Zoology.* Vol. XXV.)

1898.   Thiele, J. " Studien über pazifische Spongien." I. Heft. (*Zoologica.* Heft 24. Stuttgart.)

1899.   Thiele, J. " Studien über pazifische Spongien." II. Heft. (*Zoologica.* Heft 24^II. Stuttgart.)

1900.   Thiele, J. " Kieselschwämme von Ternate." I. (*Abh. Senckenberg. nat. Ges. Frankfurt a.M.* Bd. XXV.)

1903.   Thiele, J. " Kieselschwämme von Ternate." II. (*Abh. Senckenberg. nat. Ges. Frankfurt a.M.* Bd. XXV.)

1905.   Thiele, J. " Die Kiesel- und Hornschwämme der Sammlung Plate." (*Zool. Jahrb.* Supp. VI. Bd. III.)

1892.   Topsent, E. " Contribution à l'étude des Spongiaires de l'Atlantique Nord." (*Camp. Scient. du Prince de Monaco.* Fasc. II.)

1892 *bis.* Topsent, E.   " Diagnoses d'Éponges nouvelles de la Méditerranée et plus particulièrement
        de Banyuls."   (*Arch. de Zool. exp. et gén.*   T. X., Série 2.)
1893.   Topsent, E.   " Note sur quelques Éponges du Golfe de Tadjoura," &c.   (*Bull. Soc. Zool.
        de France.*   T. XVIII.)
1893 *bis.* Topsent, E.   " Mission scientifique de M. Ch. Alluaud aux Iles Séchelles (Mars-Mai, 1892),
        Spongiaires."   (*Bull. Soc. Zool. de France.*   T. XVIII.)
1897.   Topsent, E.   " Spongiaires de la Baie d'Amboine."   (*Revue Suisse Zool. &c.*   T. IV.)
1900.   Topsent, E.   " Étude monographique des Spongiaires de France.   III.   Monaxonida
        (Hadromérina)."   (*Arch. de Zool. exp. et gén.*   T. VIII., Série 3.)
1901.   Topsent, E.   " Spongiaires."   (*Résult. Voyage Belgica. Zool.*)
1904.   Topsent, E.   " Spongiaires des Açores."   (*Camp. Scient. du Prince de Monaco.*   Fasc.
        XXV.)
1905.   Topsent, E.   " Étude sur les Dendroceratida."   (*Arch. de Zool. exp. et gén.*   T. III.,
        Série 4).
1906.   Topsent, E.   " Éponges recueillies par M. Ch. Gravier dans la Mer Rouge."   (*Bull. Mus.
        Hist. Nat. Paris,* 1906.)
1911.   Vosmaer, G. C. J.   " The Porifera of the Siboga Expedition.   II.   The Genus Spirastrella."
        (*Siboga Expedition.*   Monographie VIa'.)
1902.   Vosmaer, G. C. J., and Vernhout, J. H.   " The Porifera of the Siboga Expedition.   I.   The
        Genus Placospongia."   (*Siboga Expedition.*   Monographie VI.)
1881.   Wright, E. P.   " On a New Genus and Species of Sponge with supposed Heteromorphic
        Zooids."   (*Trans. Royal Irish Acad. Science,* Vol. XXVIII.)

DESCRIPTIONS OF PLATES.

PLATE I.

Figs. 1a, 1b.   *Chondrilla agglutinans* n. sp. (R.N. V. 1).

Fig. 1a.   Spherasters from cortex.   × 650.
„   1b.   Oxyasters from choanosome.   × 650.

Figs. 2a–2c.   *Tetilla pilula* n. sp. (R.N. XXXV. 8 *b*)

Fig. 2a.   Outer part of surface-brush of megascleres, partially separated by teasing.   × 280.
„   2b.   Cladome of anamonæne.   × 860.
„   2c.   Sigmata.   × 860.

Figs. 3a–3d.   *Tetilla barodensis* n. sp. (R.N. XXIII. 8).

Fig. 3a.   Orthotriænes.   × 70.
„   3b.   Cladome of anatriæne.   × 290.
„   3c.   Sigmata.   × 870.
„   3d.   Trichodragma.   × 385.

Figs. 4a–4g. *Esperella plumosa* (Carter). (R.N. XXXII. 1).

Fig. 4a.	Tylostyles. × 290.	
„ 4b.	Large anisochela, front view. × 870.	
„ 4b'.	„ „ side view. × 870.	
„ 4b".	„ „ end view (small end). × 870.	
„ 4c.	Small anisochela, front view. × 870.	
„ 4c'.	„ „ side view. × 870.	
„ 4d.	Small isochela, front view. × 870.	
„ 4d'.	„ „ · side view. × 870.	
„ 4e.	Large sigma, side view. × 870.	
„ 4e'.	„ „ front view. × 870.	
„ 4f.	Small sigmata. × 870.	
„ 4g.	Toxa. × 870.	

Figs. 5a–5b⁗. *Guitarra indica* n. sp. (R.N. IV. 5).

Fig. 5a.	Styli (tornostrongyla). × 290.	
„ 5b.	Placochela, front view. × 870.	
„ 5b'.	„ ¾ front view. × 870.	
„ 5b".	„ side view. × 870.	
„ 5 b'''–5b⁗. „	developmental stages, side views. × 870.	

Figs. 6 a–6 e. *Psammochela elegans* n. gen. et sp. (R.N. XVIII. 3).

Fig. 6a.	Styli. × 290.	
„ 6b.	Larger isochela, front view. × 870.	
„ 6b'.	„ „ side view. × 870.	
„ 6c.	Small isochela, front view. × 870.	
„ 6c'.	„ „ side view. × 870.	
„ 6d.	Larger sigma, side view. × 870.	
„ 6d'.	„ „ back view. × 870.	
„ 6e.	Smaller sigmata. × 870.	

Figs. 7a–7b'. *Echinodictyum gorgonioides* n. sp. (R.N. XXI.).

Fig. 7a.	Tornotoxeote. × 380.	
„ 7b, 7b'.	Acanthostyli. × 380.	

Figs. 8a–8b. *Bubaris radiata* n. sp. (R.N. III. 7a).

Figs. 8a–8a'''.	Styli of various forms. × 90.	
Fig. 8b.	Strongyla of various forms. × 90.	

Figs. 9a, 9b. *Polymastia gemmipara* n sp. (R.N. IV. 7).

Fig. 9a.	Subtylostyle from skeleton column. × 110.	
„ 9b.	Small tylostyle. × 380.	

PLATE II.

Figs. 10a–10c. *Tetilla dactyloidea* (Carter) (R.N. II. 2).   Three specimens ; one cut in half lengthwise to show exhalant canal-system. × 1½.
Fig. 11.   *Reniera hornelli* n. sp. (R.N. IV. 4).   × 1½.
„   12.   *Reniera fibroreticulata* n. sp. (R.N. II. 8).   × 2.
„   13.   *Reniera semifibrosa* n. sp. (R.N. XXXIII. 1.).   Nat. size.
„   14a.   *Halichondria reticulata* Baer (R.N. II. 6).   Nat. size.
„   14b.   „   „   „   (R.N. II. 11).   „
„   15.   *Siphonochalina minor* n. sp. (R.N. XXVI. 4).   Nat. size.
„   16.   *Auletta lyrata* var. *glomerata* Dendy (R.N. III. 3).   × 2.
„   17.   *Auletta elongata* var. *fruticosa* nov. (R.N. XXIII. 2).   Nat. size.

PLATE III.

Fig. 18.   *Ciocalypta dichotoma* n. sp. (R.N. IV. 21).   × 2.
„   19.   *Esperella plumosa* (Carter) (R.N. XXXII. 2).   Nat. size.
„   20a.   *Desmacidon minor* n. sp. (R.N. XX. 6 a).   Flabellate specimen ; surface without oscula.
Nat. size.
„   20b.   *Desmacidon minor* n. sp. (R.N. XX. 6 b).   Digitate specimen ; osculum-bearing surface.
Nat. size.
„   21.   *Guitarra indica* n. sp. (R.N. IV. 5).   Several specimens attached to the branching tube
of a polychæte worm (*Eunice*, sp.) ; the uppermost (a) showing oscula.   Nat. size.
„   22a.   *Psammochela elegans* n. gen. et sp. (R.N. IV. 20).   Specimen preserved in alcohol, with
dermal membrane remaining.   × 1½.
„   22b.   *Psammochela elegans* n. gen. et sp. (R.N. XVIII. 3).   Specimen originally preserved in
formalin ; the dermal membrane has been removed by maceration.   × 1½.

PLATE IV.

Fig. 23.   *Echinodictyum gorgonioides* n. sp. (R.N. XXI. 1.).   Nat. size.
„   24a.   *Bubaris radiata* n. sp. (R.N. III. 7 a).   Upper surface.   × 2.
„   24b.   „   „   „   Lower surface.   × 2.
„   25.   *Spirastrella vagabunda* var. *tubulodigitata* Dendy (R.N. IV. 10).   Nat. size.
„   26a.   *Polymastia gemmipara* n. sp. (R.N. IV. 7).   Viewed from above. b. buds ; p. pebble ;
sp. c. internal spicular column.   × 3.
„   26b.   *Polymastia gemmipara* n. sp. (R.N. IV. 7).   Viewed obliquely from below, showing interior.
ch. coagulated masses of choanosomal tissue attached to inner surface of cortex. Other
lettering as before.   × 3.
„   27.   *Megalopastas retiaria* n. sp. (R.N. IV. 13).   × 2.

(All the figures on Plates II–IV are reproduced from drawings by Mr. T. P. Collings, with the
exception of Fig. 23. which is based on a photograph.)

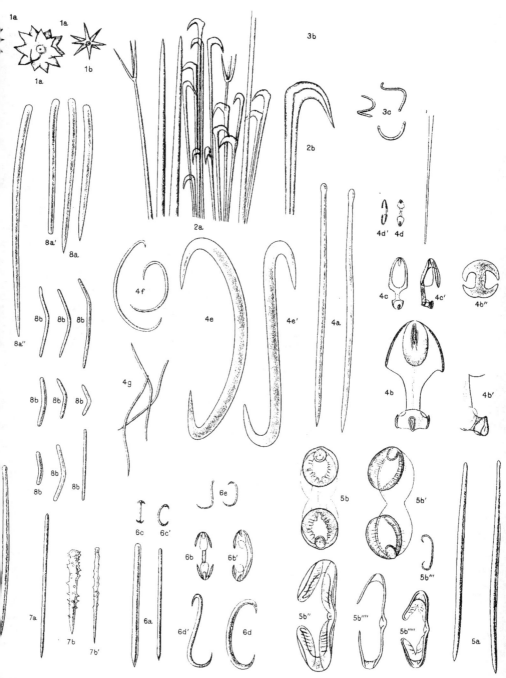

NON-CALCAREOUS SPONGES.

Cambridge University Press

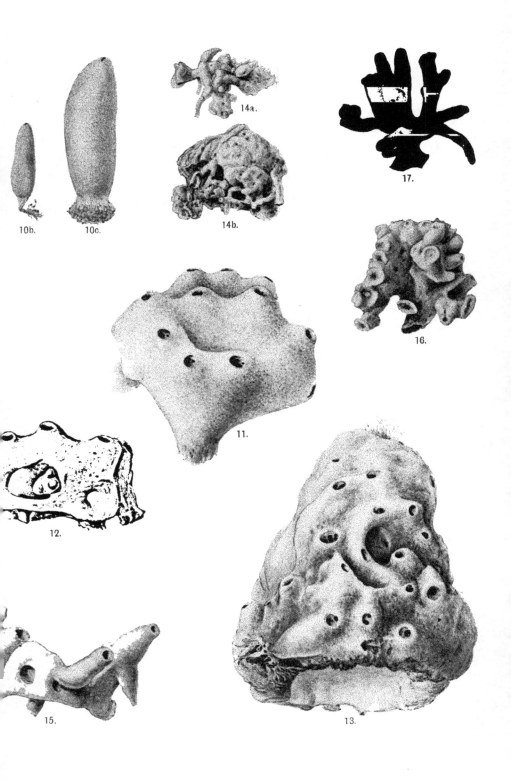

10b.

10c.

14a.

14b.

17.

16.

11.

12.

15.

13.

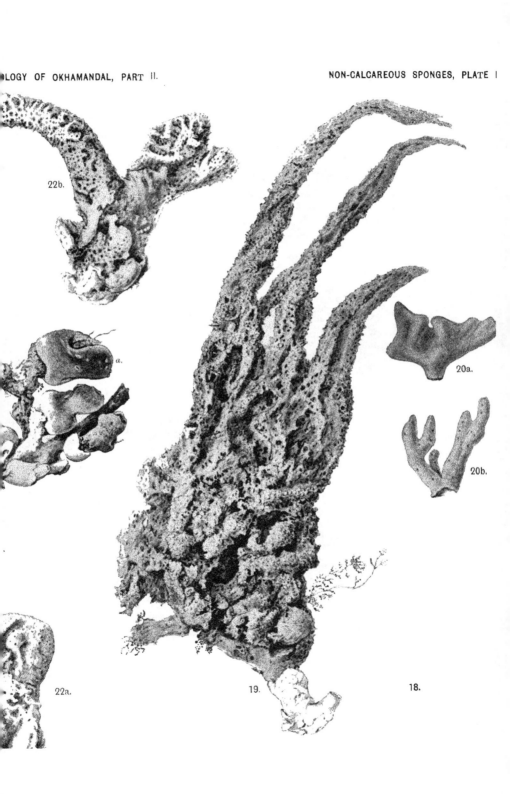

22b.

a.

20a.

20b.

22a.   19.   18.

27.

26a.

23

26b.

NON-CALCAREOUS SPONGES.

REPORT

ON THE

HYDROIDA

COLLECTED BY

MR. JAMES HORNELL

AT

OKHAMANDAL IN KATTIAWAR IN 1905-6.

BY

MISS LAURA ROSCOE THORNELY

[PREFATORY NOTE.—Of the 17 species described below, it is remarkable that whereas six only of these are found among the 43 species in the Ceylon collection described in Herdman's "Ceylon Pearl Oyster Fisheries Report," Pt. II., p. 108, as many as 10 are among the species recorded by Bale from Australian waters. When we add to these the two "Siboga" forms (*Plumularia concava* and *Monostaechas fischeri* var. *simplex*), the preponderance of Australian and Malayan species is extremely significant, constituting indeed 12 out of the 17 species represented.

Four of the six species common to Kattiawar and Ceylon are also common to Australia, and it is reasonable to expect that when the Gulf of Mannar hydroid fauna is completely investigated the remaining six species common to Kattiawar and Australia will also be found there. The present list as it stands comprises the following 10 species recorded for the first time from Indian and Ceylon coasts, viz. :—

Hebella scandens (Bale).
Synthecium maldivense Borradaile.
Pasythea quadridentata Ell. & Sol.
Thuiaria fenestrata Bale.
Monostaechas fischeri var. *simplex* Billard.

Plumularia badia Kirch.
    ,,   *caliculata* Bale.
    ,,   *concava* Billard.
Lytocarpus philippinus Kirch.
Halicornaria gracilicaulis (Jäderholm).

Special attention could not be given to the collection of Hydroids during the short period of my stay in Okhamandal, and many small and inconspicuous species undoubtedly were missed. To obtain such forms I know no better way . than to suspend, from channel buoys or from a ship's side, wire net cages containing oyster shells at a depth of a few feet below the surface.—J. HORNELL.]

HYDROIDA—SYSTEMATIC LIST.

1. Hebella scandens (Bale).

Bale.—" On Some New and Rare Hydroida in the Australian Museum Collection," *Proc. Linn. Soc. N.S.W.*, Ser. 2, Vol. III., 1888, p. 758 ; *Proc. Roy. Soc. Vict.*, 26 (N.S.), Pt. I., 1913, p. 117.

Locality :—Dwarka, 15–17 fathoms.

2. Clytia geniculata Thornely.

Thornely.—" Report on the Hydroida," in " Ceylon Pearl Oyster Fisheries," Pt. II., Suppl. Rept. No. VIII., 1904, p. 112.

A very small part of a colony.

Locality :—Off Poshetra.

3. Synthecium orthogonia (Busk).

Bale.—*Op. cit., Proc. Linn. Soc. N.S.W.*, Ser. 2, Vol. III., 1888, p. 767.

The tip of a colony most probably belongs to this species. The hydrothecæ on the lower portion of the stem are semi-alternate.

Locality :—Dwarka.

4. Synthecium maldivense Borradaile.

Borradaile.—" Fauna and Geography of the Maldive and Laccadive Archipelagoes," Vol. II., Pt. IV., p. 841.

The fragments belonging to the top portions of colonies of this species do not exceed 1½ inch in length. The branches do not show so acute an angle with the stem as in Borradaile's figure, Plate LXIX. Fig. 5a. The description of the hydrothecæ corresponds with this specimen ; the membranous hood I at first overlooked, thinking the orifice had two lateral teeth. There are several gonothecæ present on stems and branches, mostly protruding from hydrothecæ, but some are situated on the stem just below these. The grouping of hydrothecæ is remarkable, the upper four being united as in *Pasythea*. There are usually a semi-alternate pair and one in the axil of a branch to each internode. There are some nodulous internodes as described for *Thuiaria fabricii* (Levinsen).[1] There is a general

---

[1] Nutting, " American Hydroids," 1904, p. 71.

resemblance between this species and *Thuiaria tubulifera*, Marktanner-Turneretscher, as described by Clarke.[1]

Locality :—Kiu, low water.

5. Sertularia rugosissima Thornely.

Thornely.—" Hydroida " in " Ceylon Pearl Oyster Fisheries," Pt. II., Suppl. Rept. No. VIII., 1904, p. 118.

Growing in quantities over a coralline. This species appears to be identical with *S. hupferi* discovered by Broch.[2]

Locality :—Dwarka.

6. Sertularia tenuis Bale.

Bale.—*Cat. Aust. Hydroid Zoophytes*, 1884, p. 82.

Localities :—Kiu ; off Beyt.

7. Pasythea quadridentata Ellis & Sol.

Bale.—*Cat. Aust. Hydroid Zoophytes*, 1884, p. 112.

There are usually four pairs of zooecia to an internode, but sometimes only one or three pairs.

Locality :—Dwarka, 15–17 fathoms.

8. Idia pristis Lamouroux.

Bale.—*Cat. Aust. Hydroid Zoophytes*, 1884, p. 113.

Localities :—West of Beyt Island, 3–4 fathoms ; off Dwarka, 15–17 fathoms ; off Poshetra Head, 7 fathoms ; Adatra, 7 fathoms.

A widely-distributed species recorded from Torres Strait, Brazil, and the Persian Gulf.

9. Thuiaria fenestrata Bale.

Bale.—*Cat. Aust. Hydroid Zoophytes*, 1884, p. 116.

Good colonies growing on barnacles in seven fathoms of water.

Localities :—North of Poshetra, $3\frac{3}{4}$–4 fathoms ; off Poshetra Head, 7 fathoms.

10. Monostaechas fischeri Nutting (1905), variety simplex Billard.

Billard.—" Siboga Expedition," 1913, p. 16.

Localities :—Rupan Bandar ; Kutchegad.

11. Plumularia badia Kirchenpauer.

Kirchenpauer.—*Abh. ver. Hamb.* VI. ; Bale.—*Cat. Aust. Hydroid Zoophytes*, 1884, p. 128.

Locality :—Off Beyt.

---

[1] *Mem. Mus. Comp. Zool. Harvard*, Vol. XXXV. No. 1 ; VIII. "The Hydroids," 1907, p. 14.

[2] " Kenntnis der Meeresfauna West Africas," 1914, p. 35.

12. Plumularia buskii Bale.

Bale.—*Cat. Aust. Hydroid Zoophytes*, 1884, p. 125.

Locality :—Kiu.

13. Plumularia caliculata Bale.

Bale.—" On some New and Rare Hydroida in the Australian Museum Collection,"
Proc. Linn. Soc. N.S.W., Ser. 2, Vol. III., 1888, p. 780.

Locality :—Chindi Reef, 6–10 fathoms.

14. Plumularia concava Billard.

Billard.—" Siboga Expedition," 1913, p. 19.

The specimen is only half an inch in height and has no gonothecæ but is otherwise in correspondence with this species.

Locality :—Mangunda Reef, off Poshetra.

15. Lytocarpus philippinus Kirch.

Bale.—" On some New and Rare Hydroida in the Australian Museum Collection,"
Proc. Linn. Soc. N.S.W., Ser. 2, Vol. III., 1888, p. 786.

A broken colony showing however the fasciculated stem, one sided growth of the branches and hydrothecæ as described by Bale.

Locality :—Adatra.

16. Lytocarpus plumosus Thornely.

Thornely.—" Hydroida " in " Ceylon Pearl Oyster Fisheries," Pt. II., Suppl. Rept. No. VIII., p. 124.

Locality :—Channel west of Beyt Island, 3–4 fathoms.

17. Halicornaria gracilicaulis (Jäderholm).

Jäderholm.—" Aussereuropaische Hydroiden in Schwedischen Reichsmuseum," *Ark. Zool.*, Bd. I. (1903), p. 299, Taf. 14, XIV., Figs. 3–4.

Billard.—"Hydroides de Madagascar et du sud-est de l'Afrique," *Archives de Zoologie Expérimentale et Générale*, IV., Tome VII., p. 364, 1907.

There are some broken pieces, three of which belong to one colony and put together reach to a height of 6 inches and correspond with Billard's description, but there are no gonangia to help in the identification. The hydrothecæ are very delicate, having a plain rim with a point on either side. The mesial sarcotheca is always short, not protruding further than the rim.

Localities :—West of Beyt Island, 3–4 fathoms ; 3 miles W.N.W. of Samiani lighthouse ; Rupan Bandar : Kutchegad : Kiu, low water.

NOTES ON

SOME JELLY-FISHES FROM OKHAMANDAL IN KATTIAWAR

COLLECTED BY

MR. JAMES HORNELL IN 1904-5,

BY

EDWARD T. BROWNE, M.A., F.L.S.

SOME years ago Mr. James Hornell very kindly sent to me four jars of unsorted plankton and a jar containing specimens of *Cassiopea*. The plankton was collected off the coast of Okhamandal, at the entrance to the Gulf of Kutch, on the north-west coast of India, during December, 1905, and January, 1906.

The plankton was thoroughly searched for jelly-fishes, and as there were no records of their occurrence along that part of the Indian coast for several hundred miles any results would be of interest on the geographical distribution of species.

It soon became evident that the collection was more or less a failure ; at least, from my point of view. There was a decided scarcity of jelly-fishes, and an unreasonable amount of very fine mud in the jars, so fine that it took a whole day to settle. The mud not only clung tight to, but clogged up and spoilt everything in the plankton. The scarcity of jelly-fishes may be due to the collecting having been done at the wrong season of the year, or at stations too close inshore.[1]

[1] Possibly this paucity was due to the fact that the time of collection was coincident with the period when the temperature of the surface water off the Kattiawar coast is at its lowest, as this factor would be likely to influence reproduction adversely, especially in the case of the Hydromedusæ. In the tropics I have found that numerous groups of animals have two well-marked reproductive maxima in the year, and in the majority of cases these are roughly coincident with the equinoxes. Such seasonal periodicity is the most probable reason for the scarcity of Hydromedusæ at the time I visited Kattiawar. In contrast with this, the abundance of the Hydroid fauna characteristic of the littoral waters is notable. Judging from my experience elsewhere in India, I am inclined to think that March and April together with September and October are the months when plankton is most likely to be specially abundant both in species and in bulk in the Gulf of Kutch. Faunistic investigations should give the richest results at these seasons.—J. H.

The Anthomedusæ were not represented at all, and the Leptomedusæ had one species :—an *Irene*, but the specimens were too dilapidated for the determination of its species. The Trachomedusæ were represented by two species :—*Amphogona apsteini* and *Liriope* sp., which latter was fairly common, but the specimens were spoilt by the mud. The Narcomedusæ consisted solely of the well-known and widely-distributed *Solmundella bitentaculata*. The Siphonophora belonged to two species, namely *Diphyopsis chamissonis* and *Bassia bassensis* (Quoy et Gaimard). The identification of *Bassia bassensis* rests on the finding of four bracts belonging to its Eudoxid.

TRACHOMEDUSÆ.

Amphogona apsteini (Vanhöffen, 1902.)

Amphogona apsteini, Browne, 1904. "The Fauna and Geography of the Maldive and Laccadive Archipelagoes," Vol. 2, p. 740, Pl. 54, fig. 5 ; Pl. 56, fig. 1 ; Pl. 57, figs. 10–15.

Amphogona apsteini, Bigelow, 1909. *Mem. Mus. Comp. Zool. Harvard College*, p. 126, Pl. 2, figs. 1–2 ; Pl. 34, figs. 12–15 ; Pl. 45, fig. 10.

Amphogona apsteini, Mayer, 1910, "Medusæ of the World," p. 405, text fig. 257.

A single specimen was taken in December, 1905, off the coast between Rupan Bundar and Kutchegad.

The umbrella is conical, about 3 mm. in height and 4 mm. in width. The stomach is on a short peduncle, and its mouth has four lips. There are eight radial canals, and upon each one there are traces or remains of gonads which are situated about three-quarters the distance down the canals. The tentacles are all broken off, but their stumps indicate that there were twelve to fifteen in each octant. No sense-organs could be found.

The umbrella of this specimen is rather more highly arched, being higher in proportion to its width, than usually recorded, and its tentacles are more numerous, but I have no reasonable grounds for assuming it to be a new species.

Amphogona apsteini has been previously recorded from the Maldives, from Sumatra, and from Acapulco Harbour in Mexico.

NARCOMEDUSÆ.

Solmundella bitentaculata (Quoy et Gaimard, 1833.)

Solmundella bitentaculata, Browne, 1904. "The Fauna and Geography of the Maldive and Laccadive Archipelagoes," Vol. 2, p. 741, Pl. 56, fig. 3.

Solmundella bitentaculata, Browne, 1905, Medusæ, in "Report on the Pearl Oyster Fisheries of the Gulf of Manaar (Ceylon)," Part IV., p. 153, Pl. 4, figs. 1–6.

Solmundella bitentaculata, Mayer, 1910. "Medusæ of the World," p. 455.

There are twenty-two specimens from off the coast of Okhamandal. They were collected during December, 1905, and January, 1906. They are all in rather bad condition and very dirty. In general appearance they resemble the specimens described by me from Ceylon. The umbrella is highly arched, with rather a keel-like summit. The largest specimen measured 7 mm. in width and 6 mm. in height, and its tentacles about 30 mm. in length. There are no traces of peronial grooves in the perradii without tentacles. Owing to the condition of the specimens it was impossible to recognise sense-organs.

SIPHONOPHORA.

Diphyopsis chamissonis (Huxley).

Diphyes chamissonis, Huxley, 1859. "The Oceanic Hydrozoa," p. 36, Pl. 1' fig. 3.

Diphyopsis weberi, Lens and van Riemsdijk, 1908. "Siphonophora of the Siboga Expedition," p. 53, Pl. 8, figs. 67–68.

Diphyes chamissonis, Browne, 1904. "Fauna and Geography of the Maldive and Laccadive Archipelagoes," Vol. 2, p. 742, Pl. 54, fig. 6.

Diphyes chamissonis, Browne, 1905. "Pearl Oyster Fisheries of the Gulf of Manaar," p. 155.

Diphyopsis chamissonis, Bigelow, 1911, *Mem. Mus. Comp. Zool. Harvard College*, Vol. 38, p. 347.

This species was common along the coast of Okhamandal during December, 1905. The plankton was sent to me without any sorting out of the specimens, and I failed to find the posterior nectophore of any Diphyid in it, so that the posterior nectophore of this species still remains unknown.

Among the plankton were a number of Eudoxids which looked very much like the eudoxid of *Diphyopsis dispar*. The complete absence of any trace of the polygastric generation of *Diphyopsis dispar* in the plankton led me to suspect that these eudoxids had a connection with *Diphyopsis chamissonis*, whose eudoxid was unknown. On comparing the specimens with the eudoxids of *Diphyopsis dispar* from the Indian Ocean I was able to detect minute differences, sufficient, however, to isolate them. The description of the eudoxid of *Diphyopsis chamissonis* will be given in my forthcoming Report on the Siphonophora of the "*Sealark*" collection.

Diphyopsis chamissonis is widely distributed in the Indian and Pacific Oceans.

SCYPHOMEDUSÆ.

Cassiopea andromeda var. maldivensis, Browne.

Cassiopea andromeda var. *maldivensis*, Browne, 1905.   "The Fauna and Geography of the Maldive and Laccadive Archipelagoes," Vol. 2, p. 962.

Eleven specimens were taken at low water on the 9th of January, 1906, at Aramra, Beyt Bay, in the entrance to the Gulf of Kutch.   These specimens are from 20 mm. to 60 mm. in diameter, and in very good condition, even after having been in the original formalin solution for nine years.   They were sent from India in a glass-stoppered jar, one litre capacity, with the stopper sealed over with hard paraffin wax, which kept the jar perfectly air-tight and prevented any loss of formalin by evaporation.   I may here add that I always use a mixture of vaseline and beeswax for covering the rim of the stopper and it makes even common glass-stoppers proof against evaporation.

The specimens have lost all traces of colour and are now whitish.   They were originally of a dark greenish colour owing to the presence of "Green Cells" or Zooxanthellæ in the jelly.   The colour of the cells is due to chlorophyll, and it very slowly disappears, but at a faster rate if the specimens are kept in strong daylight.

The normal number of sense-organs for this species is sixteen, but a variation in number is very frequent.   Out of ten of these specimens only three have the normal number ; the others all show an increase, two have 24 sense-organs, three have 19 sense-organs, and two have 21 and 22 sense-organs respectively.

The number of velar lobes between the ocular lobes is also very variable, as the sense-organs are not usually at equal distances apart.   In these specimens three velar lobes appear to be the normal number between the ocular lobes, but any number up to five occurs, and occasionally two sense-organs are so close together that their ocular lobes are adjacent to one another, without an intervening velar lobe.

The outer margins of the ocular and velar lobes form a continuous and even margin, without any indentations, so that the lobation of the margin is practically absent between the deep ocular clefts in which the sense-organs lie.

The oral arms extend to the margin of the umbrella and are normally arranged in four pairs, but the number is not always constant.   In this series two specimens have nine arms, and one has seven arms.

The appendages on the oral arms are by no means constant in shape and size ; one specimen has long, tapering appendages predominating, another shows broad, band-shaped appendages.   In two specimens there is a band-shaped appendage in the centre where the arms meet, surrounded by numerous small appendages, varying in shape and size, forming a kind of rosette, but generally there are only a few large appendages near the central one.

Cassiopea has a very wide geographical range throughout the tropics, and lives in shallow bays and lagoons. It is rather tied to definite localities by having to pass through a fixed scyphistoma stage in the course of its life-history. It is also a variable medusa, and the finding of a *Cassiopea* in a new locality usually means the finding of a new race, called either a variety or species, or a doubtful association with a variety already named. The only method for satisfactory work on the species and varieties of this genus is the examination of a very large number of specimens from each locality.

REPORT

ON THE

POLYZOA

COLLECTED BY

MR. JAMES HORNELL

AT

OKHAMANDAL IN KATTIAWAR IN 1905-6.

BY

MISS LAURA ROSCOE THORNELY

[With Six Text-Figures]

[PREFATORY NOTE.—The following report deals with a collection made in the main by means of dredging off the western coast of Okhamandal between Dwarka and Samiani Point, and in the faunistically rich channel which constitutes Beyt Harbour, supplemented by shore-collecting on the reefs and islets at the entrance to the Gulf of Kutch. Forty-two species, and varieties have been identified by Miss Thornely, including one new species, *Beania regularis*, together with a new variety of *Bugula neritina*, which has been named var. *fastigiata*.

Considering that this is entirely a littoral and shallow water collection—the greatest depth dredged from being 17 fathoms—and that it was made by precisely the same methods as were employed during Prof. Herdman's pearl fishery investigations in Ceylon, it is remarkable that out of the 42 species now recorded, 17 only are common to the Ceylon and Kattiawar lists. The great majority of the remaining 25 species are new records for Indian waters.

The following table shows the species common to two or more of the four principal collections of Polyzoa described from India and the Indian Ocean, viz., from (a) Ceylon, reported upon in Vol. IV. of Herdman's "Ceylon Pearl Oyster Fisheries"; (b) various Indian localities, in the possession of the Indian Museum, described in Vol. I. Pt. III. of the *Records of the Indian Museum*; (c) the Indian Ocean, obtained during the "*Sealark*" Expedition of the Percy Sladen Trust Expedition in 1905 (*Trans. Linn. Soc.*, Vol. XV. Pt. I.); and, lastly, (d) the present one from the north-west extremity of Kattiawar. It is fortunate for the sake of uniformity of treatment that these collections have all been described by the same reporter, Miss Laura Roscoe Thornely, whom I take this opportunity of thanking most sincerely for having acceded to my request to identify the present collection.

DISTRIBUTION OF INDIAN POLYZOA

Species.	Kattiawar.	Ceylon.	"*Sealark*" Expedition.	Indian Museum Collection.
Aetea anguina .	×	×	—	—
Scrupocellaria cervicornis	×	×	—	×
Canda retiformis	×	—	×	×
Nellia oculata .	×	×	—	×
Bugula neritina var. *rubra*	×	—	×	—
Bugula neritina	—	×	×	×
Synnotum aviculare .	×	—	—	×
Thalamoporella rozieri	×	×	×	—
Steganoporella magnilabris	×	—	×	—
Retepora monilifera .	×	—	—	×
Microporella ciliata .	×	×	×	×
Lepralia gigas .	×	×	—	—
Mucronella thenardii	×	×	—	—
,, *coccinea*	×	×	×	—
Smittina trispinosa var. *bimucronata*	×	—	—	—
,, *trispinosa* .	—	×	×	×
Cellepora megasoma .	×	×	×	×
,, *albirostris* .	×	×	×	—
,, *tridenticulata*	×	—	×	—
Idmonea radians	×	—	×	—
,, *milneana*	×	×	×	—
Pherusa tubulosa	×	×	—	—
Amathia distans	×	×	—	× .
Buskia setigera	×	×	—	—
Alcyonidium mytili .	×	×	—	—
Cylindroecium giganteum	×	×[1]	×	—
	24	17+1	14	10

[1] Recorded elsewhere from Ceylon but not represented in the Herdman collection.

Waters' list of Red Sea species ("Repts. Soudanese Red Sea," *Journ. Linn. Soc.* Vol. XXXI., Nos. 205 and 207, 1909–10), viz. :—*Scrupocellaria cervicornis, Nellia oculata, Bugula neritina, Synnotum aviculare, Thalamoporella rozieri,* `Microporella ciliata,* and *Cylindroecium giganteum,* and that all these are cosmopolitan species ; the six first extend from the Atlantic Ocean (or Mediterranean) eastward to Australia, while the seventh, *Cylindroecium giganteum,* is recorded from British Columbia, Britain, and the Mediterranean and Red Seas.

The chief conclusion to be drawn from this comparison of collections appears to be that the polyzoan fauna of Indian Seas is still most imperfectly known and requires much more study before any useful deductions upon geographical distribution can be drawn.—J. HORNELL.]

SYSTEMATIC LIST—SUB-ORDER : CHEILOSTOMATA

1. **Aetea anguina** (Linn.).
Large colonies creeping over sea-weeds.
Locality :—Off Dwarka.

2. **Brettia tropica** Waters.
Waters.—"Bryozoa from Zanzibar," *Proc. Roy. Soc. London,* Sept. 1913, p. 465·
Growing in quantities on stems of the Hydroid *Idia pristis.*
Locality :—Off Poshetra Head, 7 fathoms.

3. **Catenicella buskii** Wyv. Thomson.
MacGillivray.—In McCoy's "Prodromus of the Zool. of Victoria," Decade III., 1879, p. 24.
A small fragment without ovicells.
Locality :—North coast of Beyt Island.

4. **Scrupocellaria pilosa** Busk.
Busk.—"Polyzoa," *Challenger Reports, Zool.,* Vol. X. Pt. XXX., 1884, p. 24.
A small fragment corresponding with Busk's description of the species but without a fornix.
Locality :—S.W. of Beyt Island (dredged).

5. **Scrupocellaria cervicornis** Busk.
Busk.—*Brit. Mus. Cat. Mar. Polyzoa,* Pt. I., 1852, p. 24.
The outer spine on the present specimens is forked as described by Waters ("Repts. Soudanese Red Sea," *Journ. Linn. Soc. Zool.,* Vol. XXXI., March 1909, p. 166).
Localities :—Rupan Bandar ; Kutchegad, 4–7 fathoms ; Dwarka.

6. **Scrupocellaria macandrei** Busk.

Busk.—*Brit. Mus. Cat. Mar. Polyzoa*, Pt. I., 1852, p. 24.

Localities :—Dwarka ; Kiu, low water.

7. **Canda retiformis** Pourtales.

Pourtales.—*Bull. Mus. Comp. Zool. Harvard Coll.*, Vol. I., No. 6, 1867, p. 110.
Waters.—*Proc. Roy. Soc. Lond.*, Sept. 1913, p. 479.
Locality :—Dwarka.

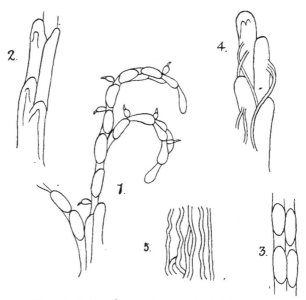

FIGS. 1—5. Bugula neritina var, fastigiata, n. var.

1. Tip of a branch ; 2. back view of the lower portion of a branch ;
3. front view of same ; 4. back view of portion of a branch showing
origins of fascicles ; 5. lower portion of a branch.

8. **Nellia oculata** Busk.

Busk.—*Brit. Mus. Cat. Mar. Polyzoa*, Pt. I., p. 18.
Localities :—Kiu, low water ; off Poshetra Head, 7 fathoms ; off Beyt.

9. **Bicellaria glabra** (Hincks).

Hincks.—*Ann. Mag. Nat. Hist.*, Ser. 5, Vol. XI., 1883, p. 195.
Locality :—Off Beyt Island.

10. **Bugula neritina** var. **rubra** Thornely.

Thornely.—*Trans. Linn. Soc. Lond.*, March 1912, p. 141.

Localities :—Off Beyt ; Dwarka, 15–17 fathoms ; Kiu, low water.

11. **Bugula neritina** var. **fastigiata** nov. (Figs. 1–5).

Zoarium with a stout, dark brown, branched, fasciculated stem rising from a fibrous root to a height of about three inches, the branches also fasciculated and dark coloured below, becoming gradually less stout and rigid until near the top where they curl gracefully over in delicate transparent tufts like those of *Bicellaria ciliata*. Zooecia biserial, stout and showing through the fascicles of the branch below, delicate and smaller in the curled upper parts ; a small avicularium below each oral aperture. Ooecia opening laterally in the usual *B. neritina* manner.

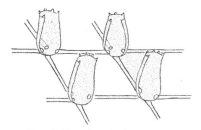

FIG. 6. **Beania regularis**, n. sp.

There are several colonies of this species which must be very beautiful when perfect, the contrast of the thick dark stems and branches and their delicate white tips being remarkable.

Localities :—Dwarka, 15–17 fathoms ; Chindi Reef, 6–10 fathoms ; Rupan Bandar ; Kutchegad.

12. **Beania regularis**, n. sp. (Fig. 6).

Zoarium loosely attached ; zooecia erect, distant, oval, open in front, widest at the base, five points on the margin above the orifice.

No lateral spines, no avicularia, connecting tubes proceeding from the lower portion of the zooecium.

Locality :—Dwarka.

13. **Synnotum aviculare** (Pieper).

Hincks.—*Ann. Mag. Nat. Hist.*, Ser. 5, Vol. XVII., p. 257.

Localities :—Dwarka, deep water ; off Beyt Island.

14. **Membranipora serrata** MacGillivray.

MacGillivray.—" Descr. Aust. Polyzoa," *Trans. Roy. Soc. Vict.*, Vol. IX.,
Nov. 1868, p. 6 ; McCoy.—*Prod. Zool. Victoria*, Decade XIII., p. 105.
A small broken colony looking like a *Flustra*.
Locality :—Off Dwarka, 15–17 fathoms.

15. **Membranipora curvirostris** Hincks.

Hincks.—*Ann. Mag. Nat. Hist.*, Ser. 3, Vol. IX., p. 29; *Brit. Mar. Polyzoa*, 1880,
p. 153.
Growing on a *Retepora*.
Locality :—Dwarka.

16. **Thalamoporella rozieri** (Andouin).

Hincks.—*Ann. Mag. Nat. Hist.*, Ser. 5, Vol. XIX., p. 164.
Some large bilaminate foliaceous pieces and also some unilaminate pieces
clinging to an *Amathia*.
Localities :—S.W. of Beyt Island ; channel west side of Beyt Island, 3–4 fathoms ;
Kiu, low water ; south of Chindi Reef, 6–10 fathoms ; off Dwarka, 15–17 fathoms.

17. **Steganoporella magnilabris** (Busk).

Busk.—" Polyzoa," *Challenger Repts. Zool.*, Vol. X. Pt. XXX., p. 75.
Localities :—Channel south-west of Beyt Island ; 3 miles W.N.W. of Samiani light-
house, 17 fathoms.

18. **Cellaria gracilis** Busk.

Busk.—*Brit. Mus. Cat. Mar. Polyzoa*, Pt. I., 1852, p. 17.
There are some good colonies growing on *Pherusa*.
Localities :—Dwarka, 15–17 fathoms ; off Poshetra, 7 fathoms.

19. **Retepora monilifera** MacGillivray.

MacGillivray.—*Trans. Phil. Inst. Vict.*, Vol. IV. Pt. 2, p. 168.
McCoy.—*Prod. Zool. Victoria*, Decade X., Vol. I. p. 20.
Some good colonies, several having a pair of long curved jointed spines one
on either side of the orifice.
Localities :—Poshetra, $3\frac{3}{4}$–4 fathoms ; south of Chindi Reef, 6–10 fathoms ; Rupan
Bandar and Kutchegad ; off Dwarka.

20. **Microporella ciliata** (Pallas).

Hincks.—*Brit. Mar. Polyzoa*, 1880, p. 206.
Growing on *Pherusa tubulosa*.
Locality :—Off Poshetra Head, 7 fathoms.

21. **Lepralia gigas** Hincks.

Hincks.—*Ann. Mag. Nat. Hist.*, Ser. 5, Vol. XV., p. 255.

Locality :—Half-mile north of Poshetra, 3¾–4 fathoms.

22. **Lepralia japonica** Busk.

Busk.—" Polyzoa," *Challenger Repts. Zool.*, Vol. X. Pt. XXX. p. 143.

Large broken foliaceous pieces and also old colonies showing no division between zooecia.

Localities :—Off Dwarka, deep water ; Hanuman Dandi Reef, Beyt ; north of Poshetra, 3¾–4 fathoms.

23. **Mucronella thenardii** (Aud.).

Savigny.—*Zool. Égypte*, Plate X. Fig. 3.

One dried specimen only.

Locality :—Off Dwarka.

24. **Mucronella coccinea** Abild.

Hincks.—*Brit. Mar. Polyzoa*, 1880, p. 371.

Locality :—Off Dwarka.

25. **Porella compressa** Sowerby.

Sowerby.—*Brit. Miscel.* I., 1806, p. 83.

Hincks.—*Brit. Mar. Polyzoa*, 1880, p. 330.

Locality :—Poshetra, 3¾–4 fathoms.

26. **Porella concinna** var. **gracilis** Hincks.

Hincks.—*Brit. Mar. Polyzoa*, 1880, p. 324.

Encrusting *Lepralia japonica*.

Localities :—Off Dwarka and Kutchegad, deep water ; off Poshetra, 7 fathoms.

27. **Smittina trispinosa** var. **bimucronata** Hincks.

Hincks.—*Ann. Mag. Nat. Hist.*, Ser. 5, Vol. XIII., p. 362.

The Dwarka specimen differs from the others in its ooecium having an arched ridge with minute pores in the area it encloses.

Localities :—Rupan Bandar ; Kutchegad ; Poshetra, 7 fathoms ; Dwarka, deep water.

28. **Haswellia australiensis** (Haswell).

Haswell.—*Proc. Linn. Soc. N.S.W.*, Vol. V. Pt. I., 1880, p. 33.

Busk.—*Challenger Repts. Zool.*, Vol. X. Pt. XXX., p. 172.

Locality :—Off Dwarka, 15–17 fathoms.

29. **Cellepora tridenticulata** Busk.

Busk.—*Challenger Repts. Zool.*, Vol. X. Pt. XXX., p. 198.

Localities :—Rupan Bandar ; Kutchegad ; Poshetra, 3¾–4 fathoms.

30. **Cellepora megasoma** MacGillivray.

MacGillivray.—*Descr. New Pol.*, Vol. VIII., p. 10.

McCoy.—*Prod. Zool. Victoria*, Decade IV., Vol. I., p. 33.

Locality :—Off Dwarka.

31. **Cellepora bispinata** Busk.

Busk.—*Brit. Mus. Cat. Mar. Polyzoa*, Pt. II., 1854, p. 87.

There are the bases only of two spines on the one small colony present. The rostrum sometimes rises to a thin point and has no avicularium upon it. Vicarious avicularia differ in size from being very small to larger than the zooecia.

Localities :—Rupan Bandar ; Kutchegad.

32. **Cellepora albirostris** (Smitt).

Smitt.—" Floridan Bryozoa," Pt. II., p. 70, in *Vetensk. Akad. Handl.*, Vol. XI., 1872.

Busk.—*Challenger Repts. Zool.*, Vol. X. Pt. XXX., p. 193.

Localities :—South-west coast of Beyt Island ; Dwarka, deep water, growing on *Lepralia japonica ;* north of Poshetra, 3¾–4 fathoms.

SUB-ORDER : CYCLOSTOMATA.

33. **Idmonea radians** (Lamk.).

Busk.—*Brit. Mus. Cat. Mar. Polyzoa*, Pt. III., p. 11, 1875.

Locality :—Off Poshetra Head, 7 fathoms.

34. **Idmonea australis** MacGillivray.

McCoy.—*Prod. Zool. Victoria*, Decade VII., p. 30.

Locality :—Off Dwarka.

35. **Idmonea milneana** D'Orb.

D'Orb.—*Voy. Amer. Merid.*, " Polypiers," p. 20.

Busk.—*Brit. Mus. Cat. Mar. Polyzoa*, Pt. III., p. 12, 1875.

Locality :—Dwarka, 15–17 fathoms.

36. **Pustulopora deflexa** Smitt.

Smitt.—" Floridan Bryozoa," Pt. I., p. 11.

Busk.—*Challenger Repts. Zool.*, Vol. XVII. Pt. L., p. 20.

Locality :—Off Dwarka.

38. **Amathia distans** Busk.

Busk.—*Challenger Repts. Zool.*, Vol. XVII. Pt. L., p. 33.

Fragments of colonies only.

Localities :—South-west of Beyt Island ; Kiu, low water ; Poshetra, $3\frac{3}{4}$-4 fathoms ; Rupan Bandar ; Kutchegad ; off Poshetra, 7 fathoms ; channel west of Beyt Island, 3-4 fathoms ; Chindi Reef, 6-10 fathoms.

39. **Amathia connexa** Busk.

Busk.—*Challenger Repts. Zool.*, Vol. XVII. Pt. L., p. 35.

Localities :—Kiu ; Dwarka.

40. **Buskia setigera** Hincks.

Hincks.—*Journ. Linn. Soc.*, Vol. XXI., p. 127.

Growing on stems of *Idia pristis.*

Locality :—Off Poshetra Head, 7 fathoms.

41. **Alcyonidium mytili** Dalyell.

Hincks.—*Brit. Mar. Polyzoa*, 1880, p. 498.

Locality :—South-west of Beyt Island.

42. **Cylindroecium giganteum** (Busk).

Busk.—*Quart. Journ. Micro. Soc.*, Vol. IV., p. 93.

Hincks.—*Brit. Mar. Polyzoa*, 1880, p. 535.

Localities :—Dwarka, 15-17 fathoms ; off Poshetra Head, 7 fathoms.

(MS. received from author 29th July, 1915).

CPSIA information can be obtained
at www.ICGtesting.com
Printed in the USA
BVHW08*1229021018
529052BV00008B/502/P